Disclaimer

The publisher of this book is by no way associated with the National Institute of Standards and Technology (NIST). The NIST did not publish this book. It was published by 50 page publications under the public domain license.

50 Page Publications.

Book Title: Design of the HLPR ChairHome Lift Position and Rehabilitation Chair

Book Author: Roger V. Bostelman; James S. Albus

Book Abstract: This paper is a HLPR Chair (Home Lift, Position, and Rehabilitation Chair) design document that includes a brief introduction of the need for this technology and all the mechanical and electrical designs produced to date to build this technology.

Citation: NIST Interagency/Internal Report (NISTIR) - 7459

Keyword: assistive care;lift;position;rehabilitation;robotic wheelchair;transfer device

Design of the HLPR Chair
Home Lift Position and Rehabilitation Chair

Roger Bostelman
James Albus

Intelligent Systems Division
Manufacturing Engineering Laboratory
National Institute of Standards and Technology

June 19, 2007

Table Contents

Introduction..3
    Wheelchairs..3
    Patient Lift..3
    Introduction References...4
Concept..6
Prototype..9
Specifications...12
Mechanical Design...13
    HLPR Chair Component Labels...17
Modifications to Original Designs...51
Electrical Design..70
Towards Autonomous Control...81
Stability Testing...87
Prototype Cost Estimate...88

Introduction

Reference [1] says "today, approximately 10 percent of the world's population is over 60; by 2050 this proportion will have more than doubled" and "the greatest rate of increase is amongst the oldest old, people aged 85 and older." She follows by adding that this group is subject to both physical and cognitive impairments more than younger people. These facts have a profound impact on how the world will maintain the elderly independent as long as possible from caregivers. Both physical and cognitive diminishing abilities address the body and the mental process of knowing, including aspects such as awareness, perception, reasoning, intuition and judgment. Assistive technology for the mobility impaired includes the wheelchair, lift aids and other devices, all of which have been around for decades. However, the patient typically or eventually requires assistance to use the device; whether it's someone to push them in a wheelchair, to lift them from the bed to a chair or to the toilet or for guiding them through cluttered areas. With fewer caregivers and more elderly, there is a need for improving these devices to provide them independent assistance.

Wheelchairs

There has been an increasing need for wheelchairs over time. L.H.V. van der Woude [2] states that mobility is fundamental to health, social integration and individual well-being of the humans. Henceforth, mobility must be viewed as being essential to the outcome of the rehabilitation process of wheelchair dependent persons and to their successful (re-)integration into society and to a productive and active life. Thrun [3] said that, if possible, rehabilitation to relieve the dependence on the wheelchair is ideal for this type of patient to live a longer, healthier life. Van der Woude continues stating that many lower limb disabled subjects depend upon a wheelchair for their mobility. Estimated numbers for Europe and USA are respectively 2.5 million and 1.25 million. The quality of the wheelchair, the individual work capacity, the functionality of the wheelchair/user combination, and the effectiveness of the rehabilitation program do indeed determine the freedom of mobility.

Patient Lift

Just as important as wheelchairs are the lift devices and people who lift patients into wheelchairs and other seats, beds, automobiles, etc. The need for patient lift devices will also increase as generations get older. When considering if there is a need for patient lift devices, several references state the positive, for example:

- "The question is, what does it cost not to buy this equipment? A back injury can cost as much as $50,000, and that's not even including all the indirect costs. If a nursing home can buy these lifting devices for $1,000 to $2,000, and eliminate a back injury that costs tens of thousands of dollars, that's a good deal," [4]
- 1 in every 3 nurses become injured from the physical exertion put forth while moving non-ambulatory patients; costing their employers $35,000 per injured nurse. [5]
- 1 in 2 non-ambulatory patients fall to the floor and become injured when being transferred from a bed to a wheelchair. - [6]
- "Nursing and personal care facilities are a growing industry where hazards are known

and effective controls are available," said OSHA Administrator John Henshaw. "The industry also ranks among the highest in terms of injuries and illnesses, with rates about 2 1/2 times that of all other general industries..." [7]

- "Already today there are over 400,000 unfilled nursing positions causing healthcare providers across the country to close wings or risk negative outcomes. Over the coming years, the declining ratio of working age adults to elderly will further exacerbate the shortage. In 1950 there were 8 adults available to support each elder 65+, today the ratio is 5:1 and by 2020 the ratio will drop to 3 working age adults per elder person." [8]

In 2005, NIST ISD began the Healthcare Mobility Project to target this staggering healthcare issue of patient lift and mobility. ISD researchers looked at currently available technology through a survey of patient lift and mobility devices [9]. That report showed that there is need for technology that includes mobility devices that can lift and maneuver patients to other seats and technology that can provide for rehabilitation to help the patient become independent of the wheelchair.

An additional area investigated in the survey was intelligent wheelchairs. NIST has been studying intelligent mobility for the military, transportation, and the manufacturing industry for nearly 30 years through the Intelligent Control of Mobility Systems (ICMS) Program. [10] Toward a standard control system architecture and advanced 3D imaging technologies, as being researched within the ICMS Program, and applying them to intelligent wheelchairs, NIST has begun outfitting the HLPR Chair with computer controls. Although throughout the world there are or have been many research efforts in intelligent wheelchairs, including: [11, 12, 13, 14] and many others, the authors could find no sources applying standard control methods nor application of the most advanced 3D imagers prototyped today to intelligent wheelchairs. Therefore, NIST began developing the HLPR Chair [15] to investigate these specific areas of mobility, lift and rehabilitation, as well as advanced autonomous control.

This paper includes mechanical and electrical designs for the HLPR Chair in its prototype stage with designs dating back to December 2005. An initial prototype HLPR Chair concept, called RoboChair, was completed in July 2004.

Introduction References

[1] Pollack, Martha, "Intelligent Technology for Adaptive Aging" Presentation, AAAI-04 American Association for Artificial Intelligence Conference Keynote Address, 2004
[2] L.H.V. van der Woude, M.T.E. Hopman and C.H. van Kemenade, "Biomedical Aspects of Manual Wheelchair Propulsion: The State of the Art II," Volume 5, Assistive Technology Research Series, 1999, 392 pp., hardcover
[3] Thrun, Sebastian, Visit to Stanford University to discuss healthcare mobility devices, August 2006.
[4] Marras, William, "Lifting Patients Poses High Risk for Back Injuries," Ohio State University, http://researchnews.osu.edu/archive/resthome.htm. 1999.
[5] Blevins, Healthcare Statistics: Blevins Medical, Inc., http://www.patientlift.net/282164.html. 2006
[6] U.S. Bureau of Labor Statistics, from Blevins website: http://www.patientlift.net/282164.html, 1994.

[7] John Henshaw, http://www.osha.gov/SLTC/nursinghome/solutions.html, Occupational Safety and Health Administration, 2005

[8] Wasatch Digital iQ, "InTouch Health's Remote Presence Robot Used by Healthcare Experts," http://www.wasatchdigitaliq.com/parser.php?nav=article&article_id=43, Santa Barbara, CA & Salt Lake City --(Business Wire)--June 16, 2003.

[9] Bostelman, Roger; Albus, James, "Survey of Patient Mobility and Lift Technologies Toward Advancements and Standards" NISTIR #7384, 2006.

[10] NIST Intelligent Control of Mobility Systems Program website: http://www.isd.mel.nist.gov/research_areas/mobility/index.htm

[11] Kuno, Y., Murashima, T., Shimada, N., Shirai, Y., "Intelligent Wheelchair Remotely Controlled by Interactive Gestures," International Conference on Pattern Recognition, vol. 04, no. 4, p. 4672, 2000.

[12] Patel, S., Jung, S-H., Ostrowski, J., Rao, R., Taylor, C., "Sensor based door navigation for a nonholonomic vehicle," GRASP Laboratory, University of Pennsylvania, Proceedings of the 2002 IEEE International Conference on Robotics and Automation, Washington, DC, May 2002.

[13] Song W.-K.; Lee H.; Bien Z., "KAIST - KARES: Intelligent wheelchair-mounted robotic arm system using vision and force sensor," Robotics and Autonomous Systems, vol. 28, no. 1, pp. 83-94(12), 31, Publisher: Elsevier Science, July 1999.

[14] Yanco, H., Hazel, A., Peacock, A., Smith, S. and Wintermute, H. "Initial Report on Wheelesley: A Robotic Wheelchair System," Department of Computer Science, Wellesley College, 1995

[15] Bostelman, R., Albus, J., "HLPR Chair – A Service Robot for the Healthcare Industry," 3rd International Workshop on Advances in Service Robotics, Vienna, Austria, July 7, 2006

Concept

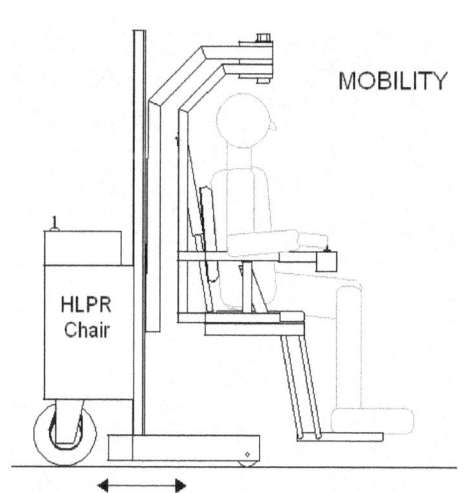
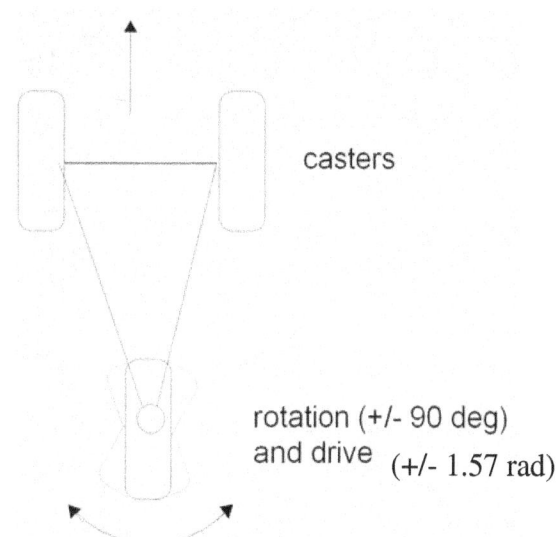

HLPR Chair is shown in the powered chair configuration with controls that the patient can use to drive the chair or a nurse/caregiver can use the controls from behind the chair to mobilize the patient.

Top view of the HLPR Chair rear steering and drive wheel and fixed front casters configuration.

(a)

Design of the HLPR Chair

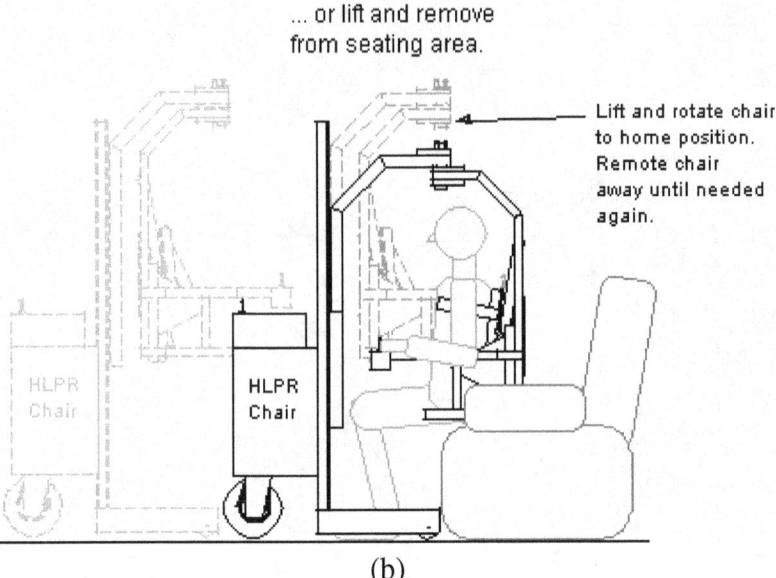

(b)

(a) HLPR Chair can be used to carry a patient to a chair or toilet, rotate the patient ready to place them on the seat, remove the footrest, raise the torso lifts and patient above the HLPR Chair seat, remove the HLPR Chair seat from beneath the patient, and place the patient on the target chair or toilet. (b) HLPR Chair can be used for patient placement on a comfortable chair. Additionally, controls can be designed to allow the patient to command using voice or remote interface to control the HLPR Chair go away until or come when needed.

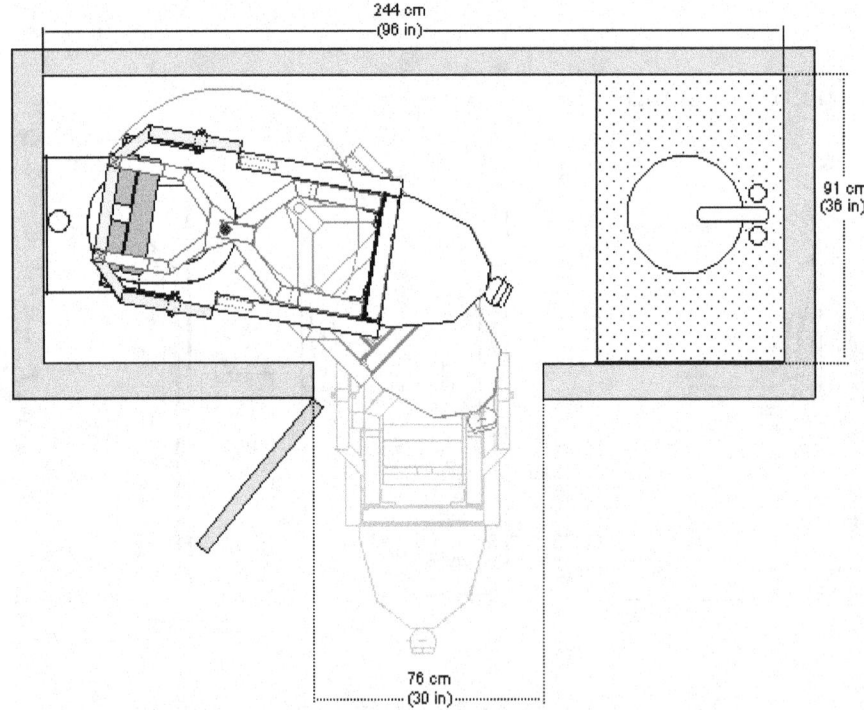

Top view showing HLPR Chair accessing a toilet in a bathroom. The red arc shows the maximum radius of seat frame rotation needed.

Design of the HLPR Chair

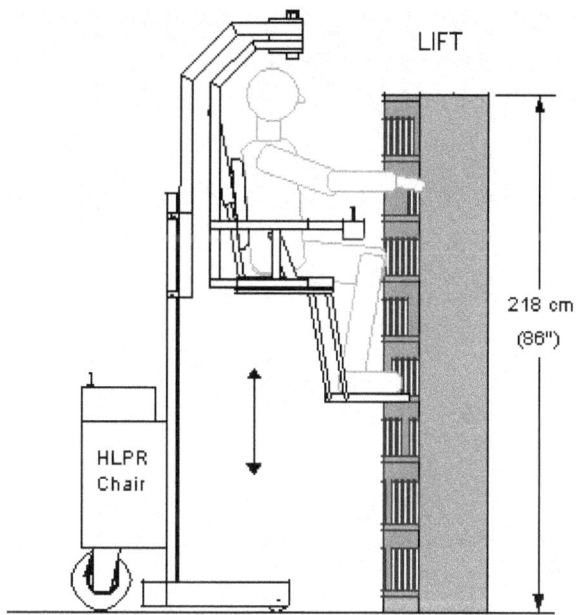

HLPR Chair allows lift, currently up to 0.9 m (36 in) above mobility configuration height.

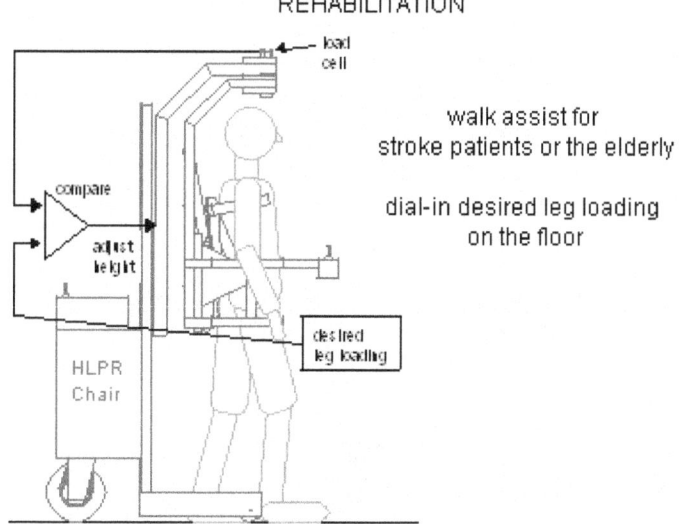

HLPR Chair can be used for patient rehabilitation and incorporate future legs load control.

Prototype

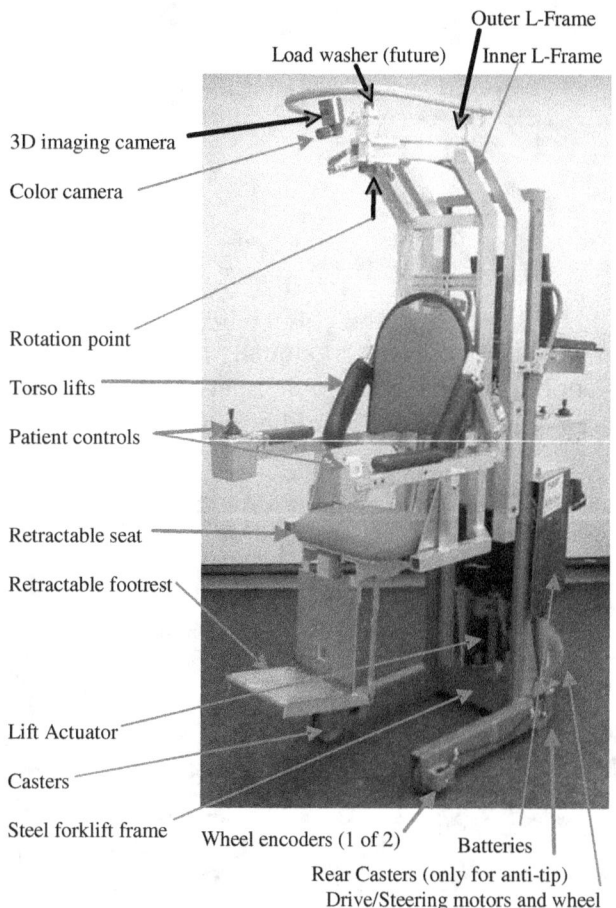

Photograph 1 of the HLPR Chair prototype.

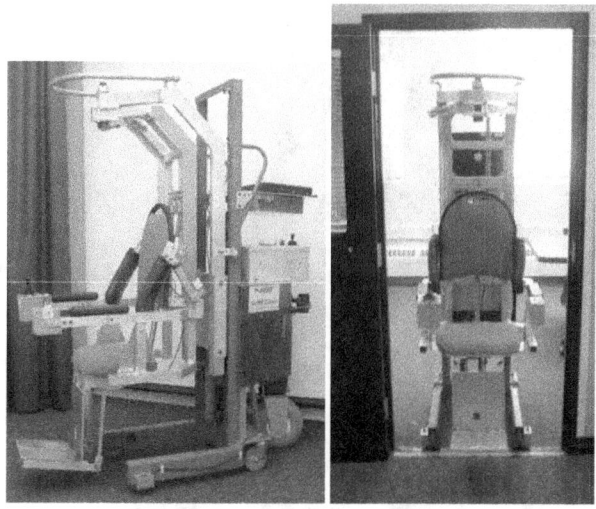

Photograph 2 (left) of the HLPR Chair in the mobility configuration showing the side view and Photograph 3 (right) front view relative to a typical doorway.

Design of the HLPR Chair

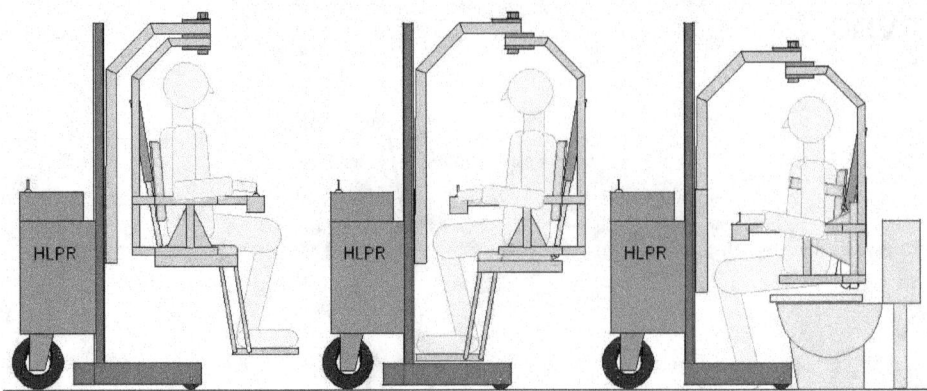

Graphic showing the concept of placing a patient onto a toilet or chair with the HLPR Chair. The patient drives to the target seat (left), manually rotates near or over the seat (middle) while the torso lifts support the patient and the seat retracts, and then is lowered onto the seat - toilet, chair or bed (right)

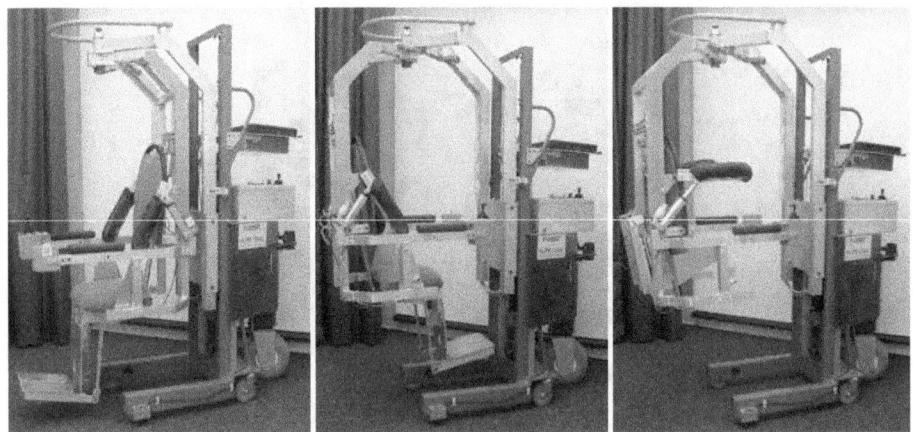

Photographs 4, 5, and 6 (lt. to rt.) of the HLPR Chair prototype in the same configurations as in the graphic above placing a patient on a toilet.

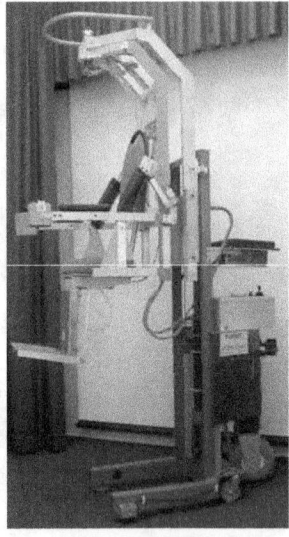

Photograph 7 of the HLPR Chair prototype shown in the patient lift position.

Design of the HLPR Chair

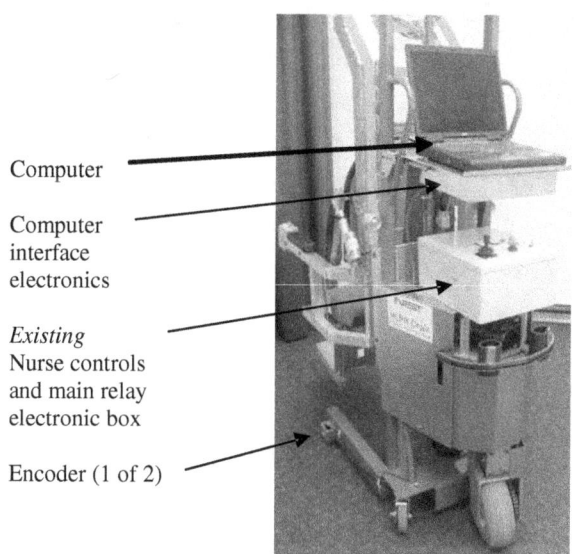

Photograph 8 of the HLPR Chair with recently added front wheel encoders, development computer and interface electronics.

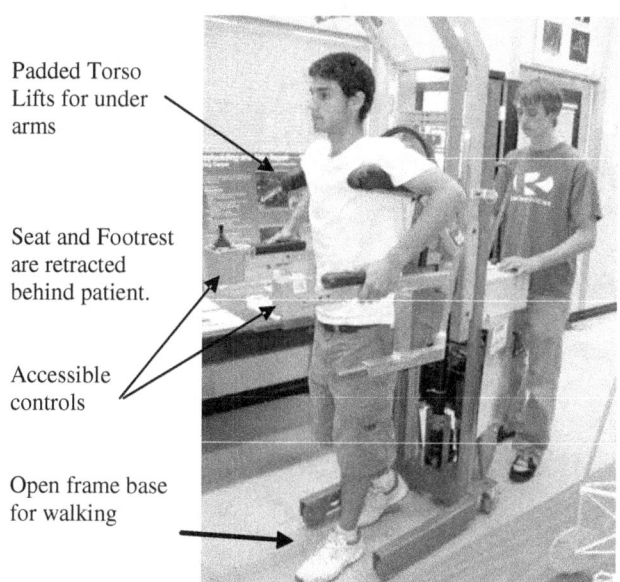

Photograph 9 of the HLPR Chair prototype in the rehabilitation/walking configuration. Summer Interns (Alex Page and Robert Vlacich) demonstrate the patient and nurse configuration as part of their official duties.

Specifications

Size:	
Mobility Configuration	145 cm long x 58 cm wide x 178 cm high
	(57" long x 23" wide x 70" high)
	with 57 cm (22 ½") seat ht. above floor
Full Lifted Configuration	145 cm long x 58 cm wide x 241 cm high
	(57" long x 23" wide x 95" high)
	with 125 cm (49") seat ht. above floor – currently *can be adjusted to lift 91 cm (36")*
Weight (unloaded)	136 kg (300 Lbs.)
Payload:	136 kg (300 Lbs.) (designed)
	91 kg (200 Lbs.) (tested to date)
Tilt	0.06 rad (10 deg)
Max. Speed	0.7 mps (28 ips)
Turning Radius	86 cm (34") centered about the rider
Chair Rotate Angle	0.5 rad (90 deg) CCW to 1 rad (180 deg) CW
Wheels:	
Rear Drive/Steer	10" diameter pneumatic
Front Caster	5" diameter solid
Ground Clearance	4.4 cm (1 ¾")
Battery	2-12Vdc dry cells (series 24V)
Per-Charge Range	unknown to date
Battery weight	11.6 kg (26 Lbs) each
Drive Train	1 motor chain drive, 1 gearmotor direct steer
Battery Chargers	two (one per battery), off-board

Mechanical Design

Manual Fork Lift Base Frame[1]

Hydraulic Lift
  Foot operated hydraulic pump makes raising the load a simple task.
  Foot pedal release results in controlled lowering to reduce the risk of accidents.
  Wheels are on swivel casters and lock into position for load lifting and lowering.
  Two forks lower to 10.8 cm (4.3 in) above the floor and raise loads up to 137 cm (54 in) to get your machinery or heavy objects at an accessible level.
  227 kg (500 Lbs) capacity.

For the HLPR Chair Design, the hydraulic piston was replaced with an electric ball screw actuator with 454 kg (1000 Lbs) lift capacity and 0.45 m (18 in) stroke. The fork lift chain design allows a 2:1 lift height for 0.9 m (36 in) fork (chair) lift with half the payload or 227 kg (500 Lbs).

---

[1] NIST does not endorse products or organizations.

Design of the HLPR Chair

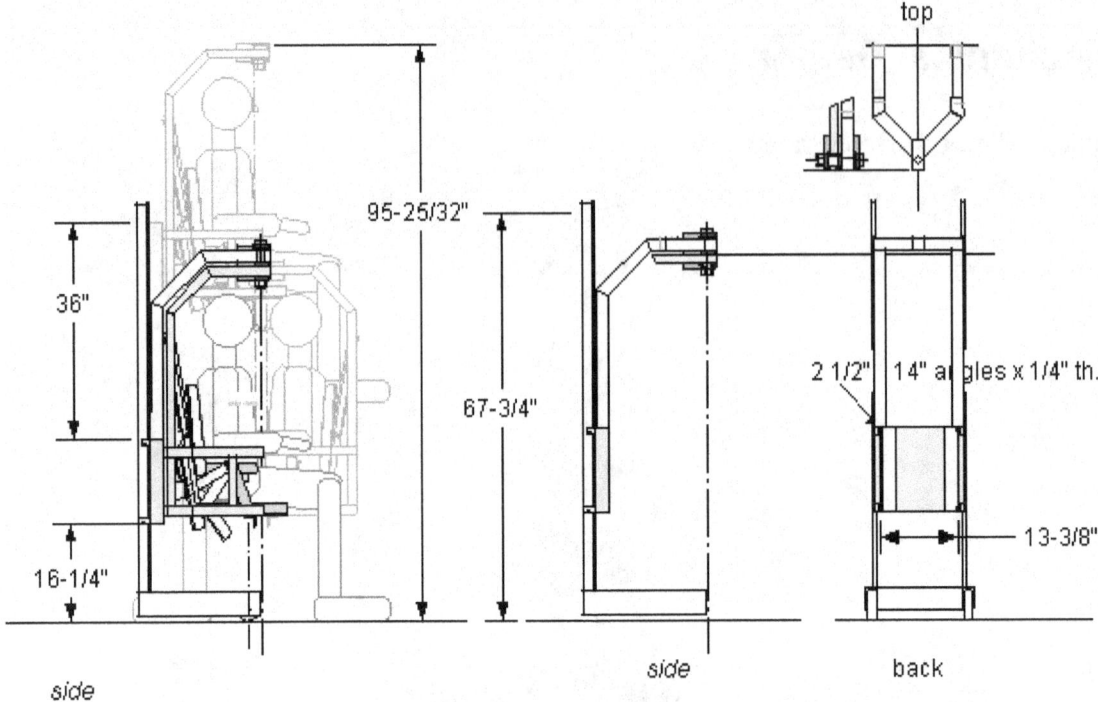

Left (side) view shows 0.9 m (36 in) base and seat frame travel and maximum system height. Center side view shows 37.4 cm (95 25/32 in). Center (side) and right (back) views show only the base frame with the forklift frame components with the seat frame removed and the rotation (top) joint view.

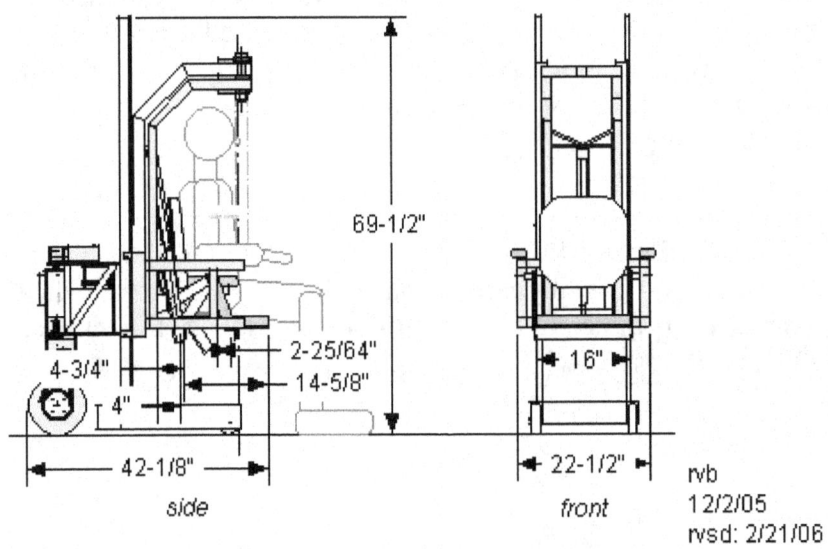

Side and front views of the HLPR Chair with minimal dimensions for length, height, and width while in the mobility configuration.

Design of the HLPR Chair

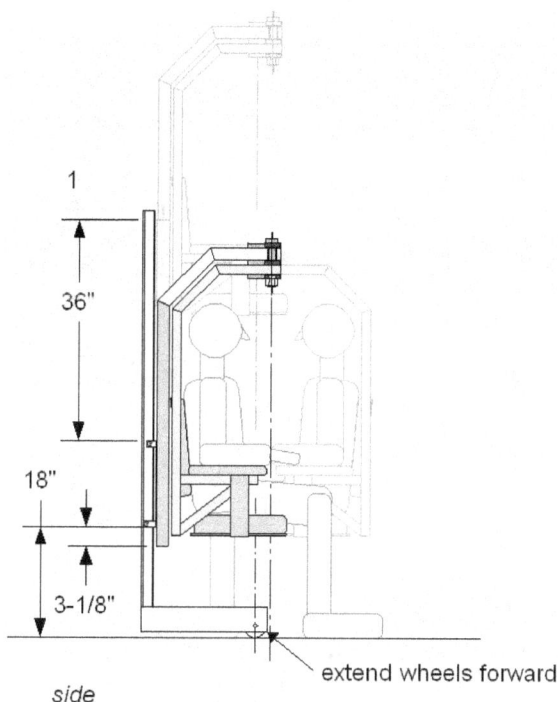

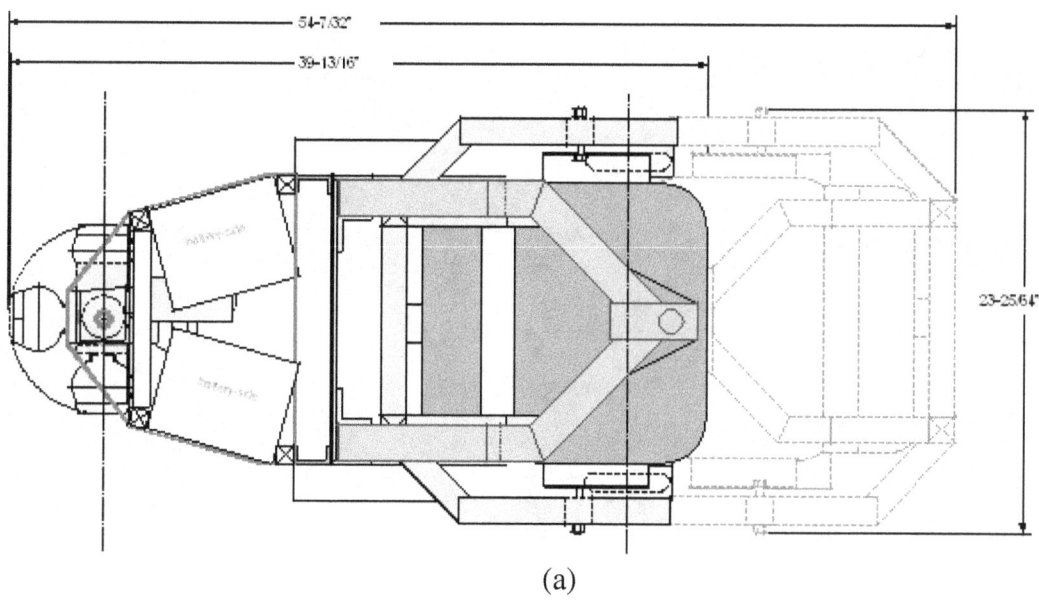

HLPR Chair side view without the rear drive system and suggested front wheel extension to allow the seat rotation point to be within the wheelbase. This modification was not added to the prototype and is not necessary with appropriate counterweight (e.g., drive and battery system) added to the rear.

(a)

Design of the HLPR Chair

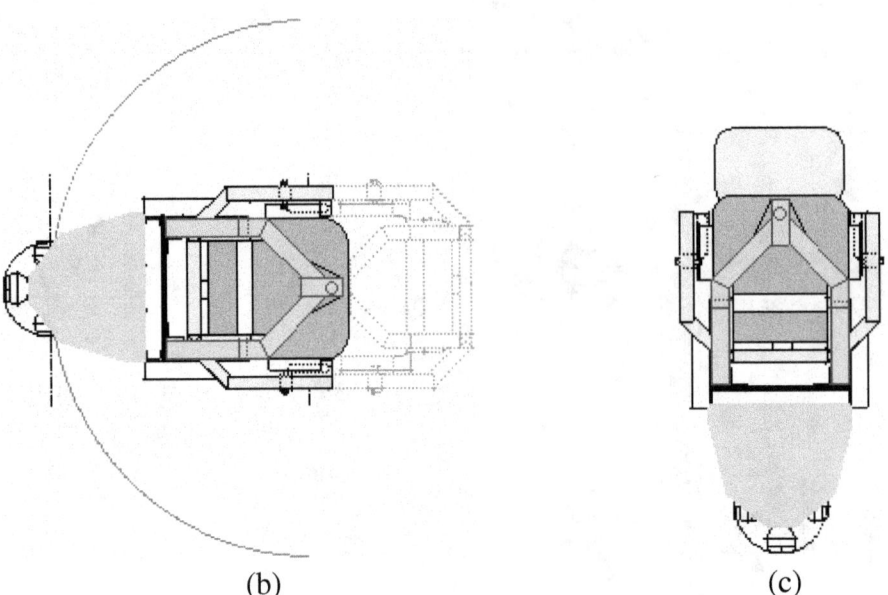

(b)　　　　　　　　　　　　(c)

Top views of the HLPR Chair showing (a) Maximum length and width, (b) minimum HLPR Chair rotation about the front casters axle center 76 cm (30 in) and (c) top view of the HLPR Chair.

HLPR Chair Component Labels
These labels refer to drawings that follow.

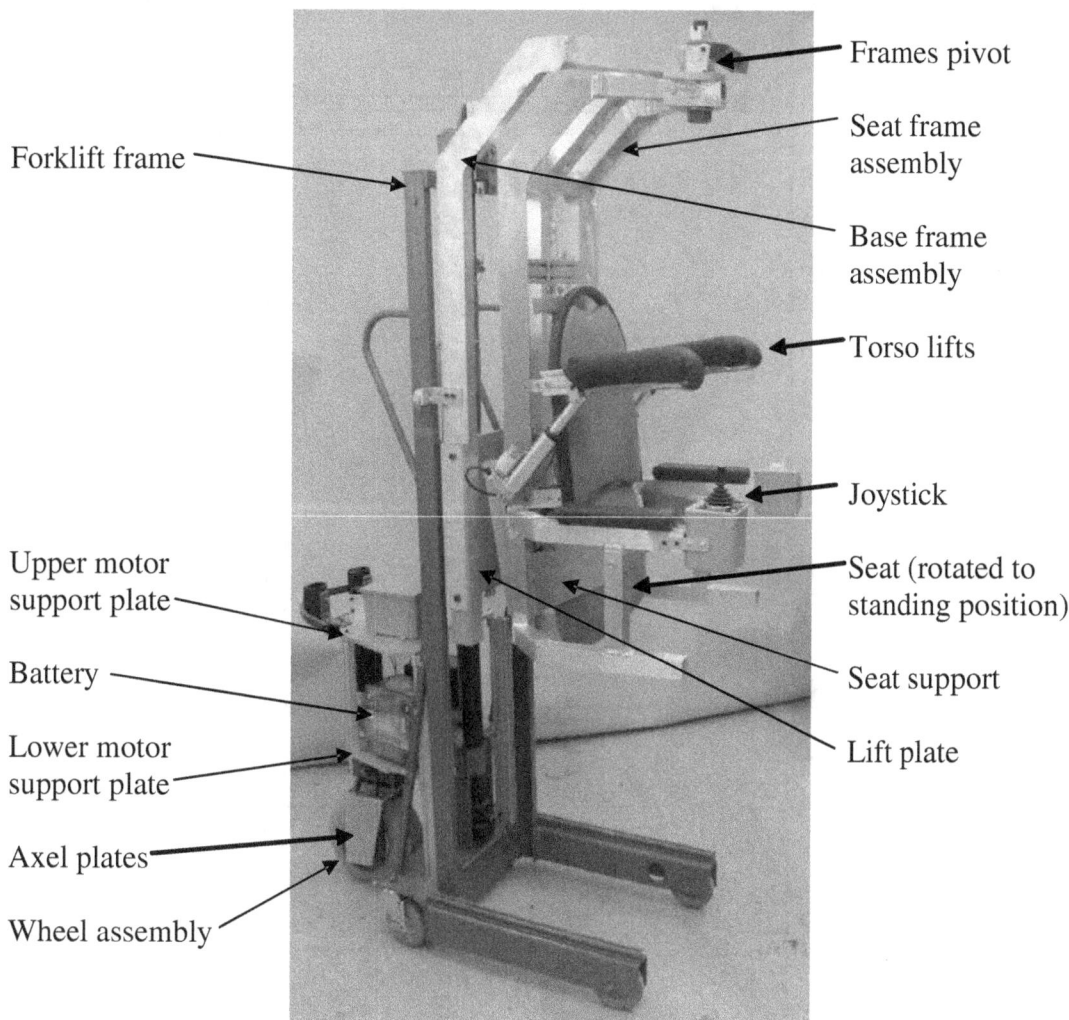

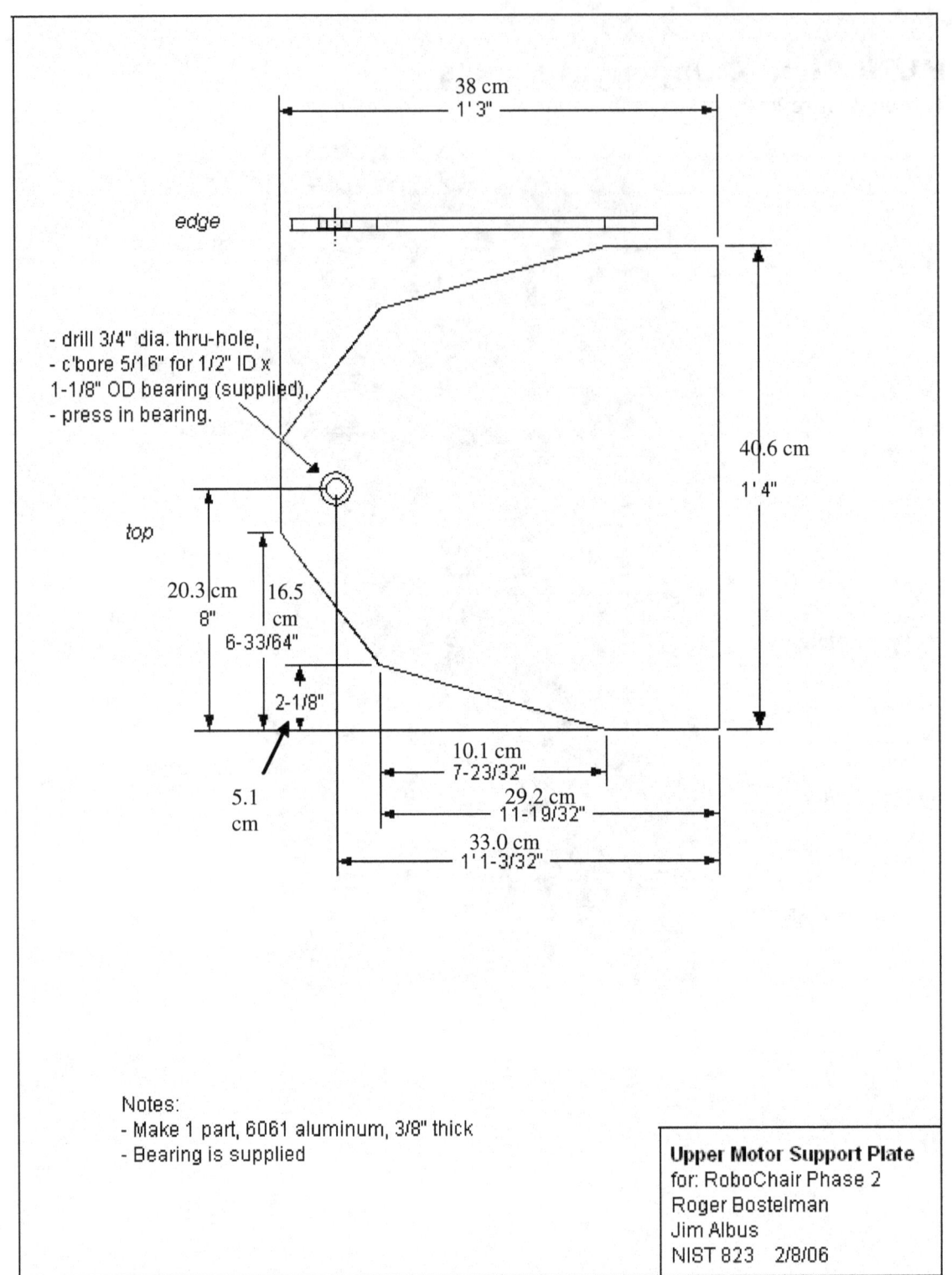

Design of the HLPR Chair

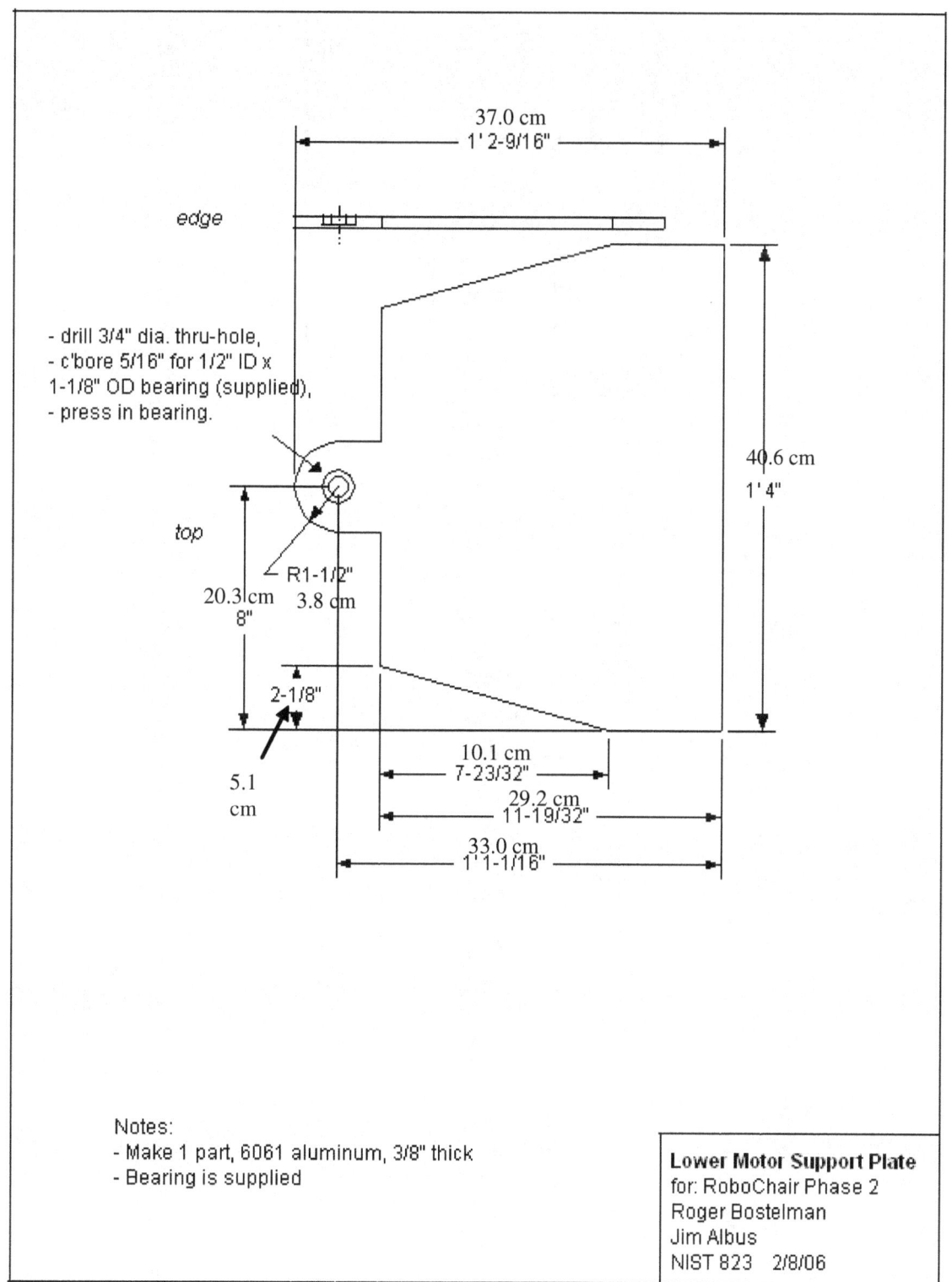

Design of the HLPR Chair

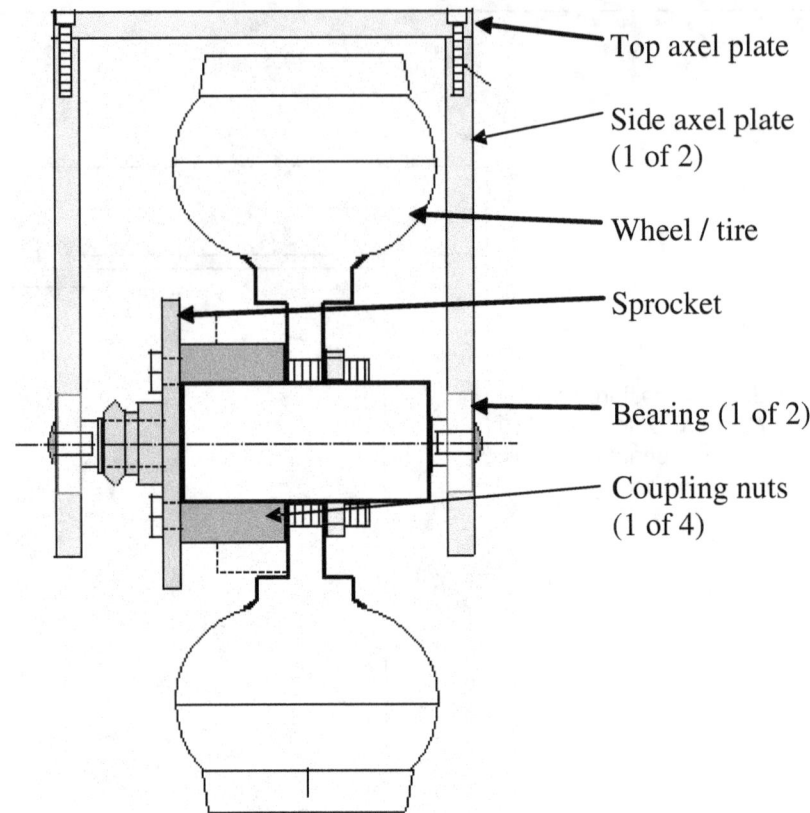

Front view of wheel assembly
(component drawings for this assembly follow)

Design of the HLPR Chair

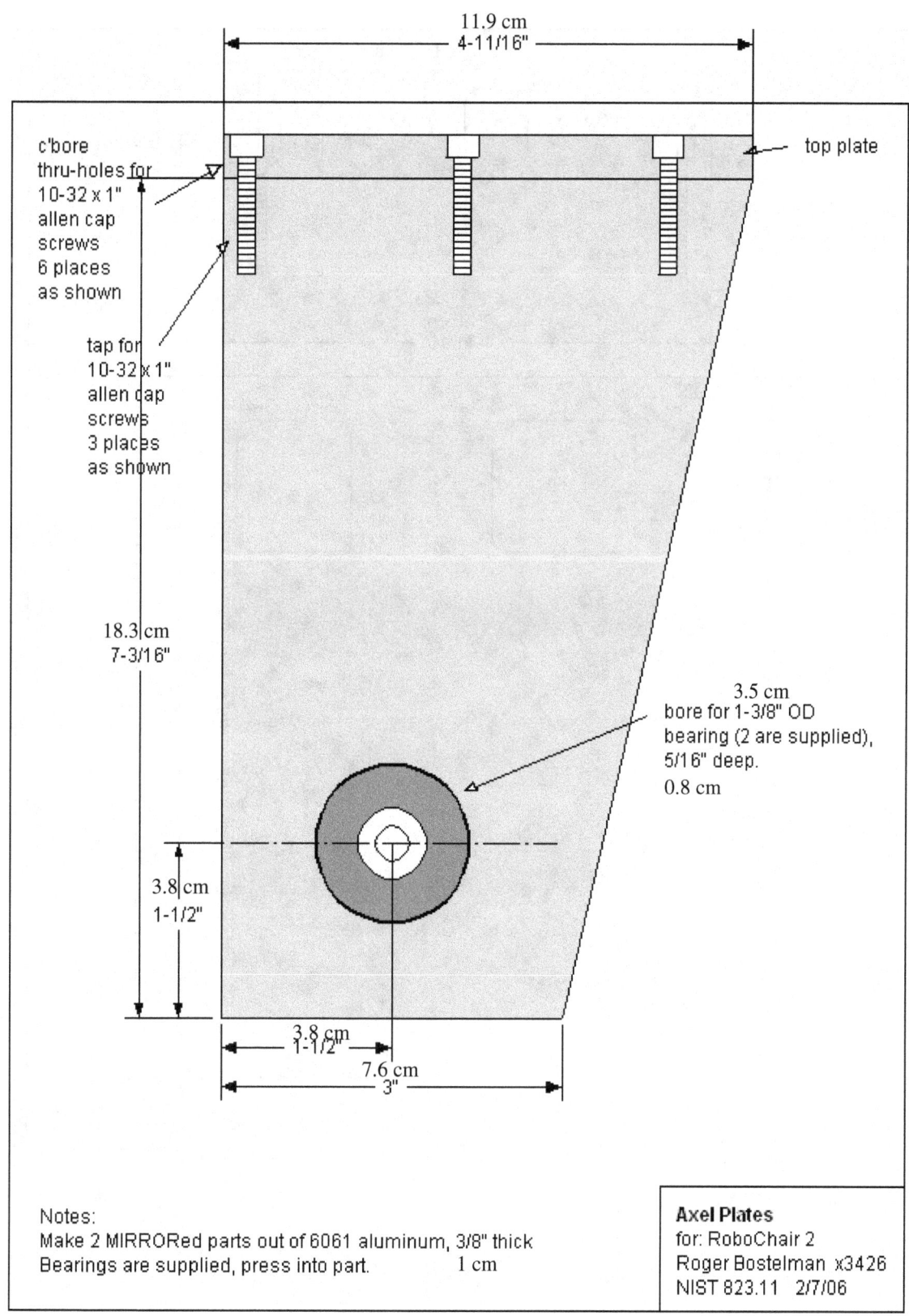

Design of the HLPR Chair

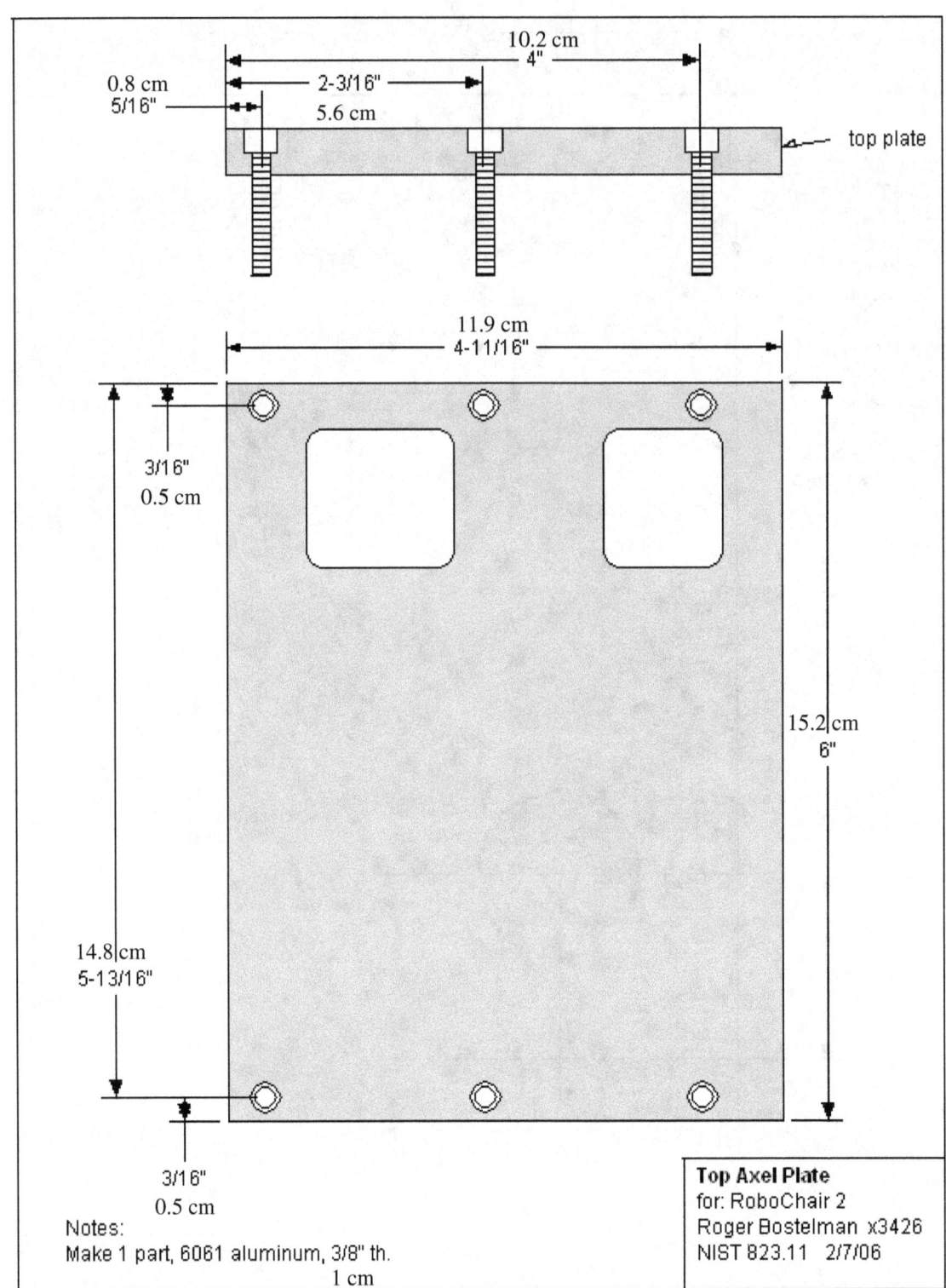

Notes:
Make 1 part, 6061 aluminum, 3/8" th.

Top Axel Plate
for: RoboChair 2
Roger Bostelman x3426
NIST 823.11 2/7/06

Design of the HLPR Chair

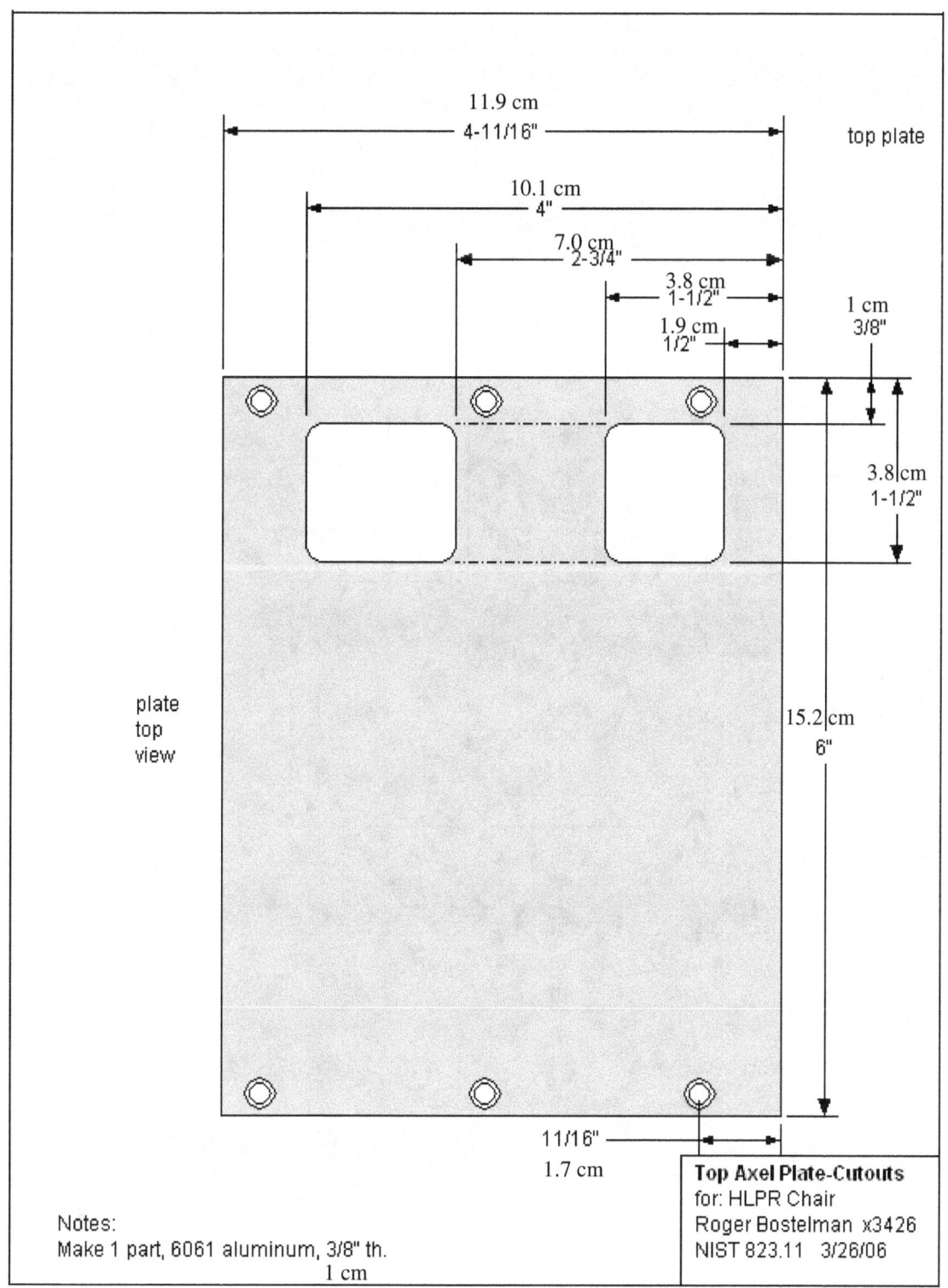

Design of the HLPR Chair

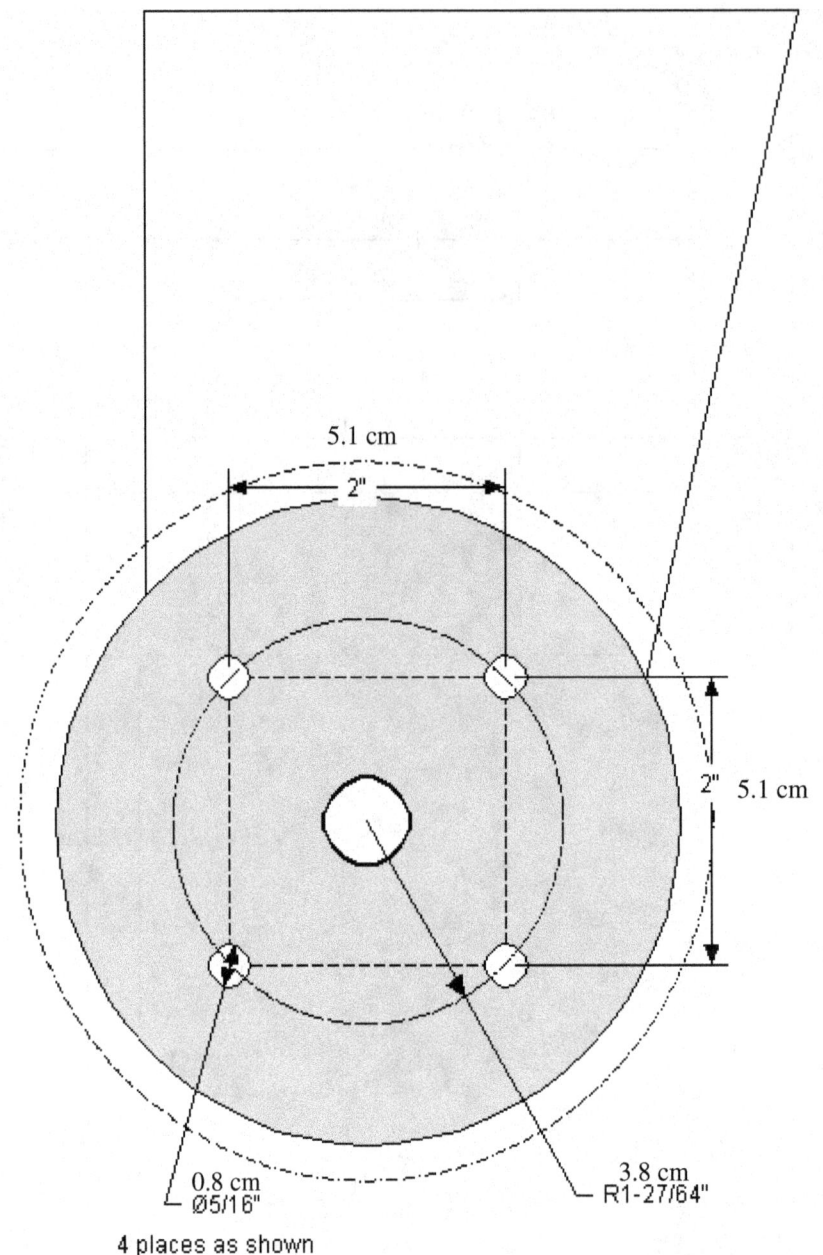

Notes:
drill 4 holes as shown on drawing

Wheel Drive Sprocket
for: RoboChair 2
Roger Bostelman x3426
NIST 823.11 2/7/06, 2/22/06

Design of the HLPR Chair

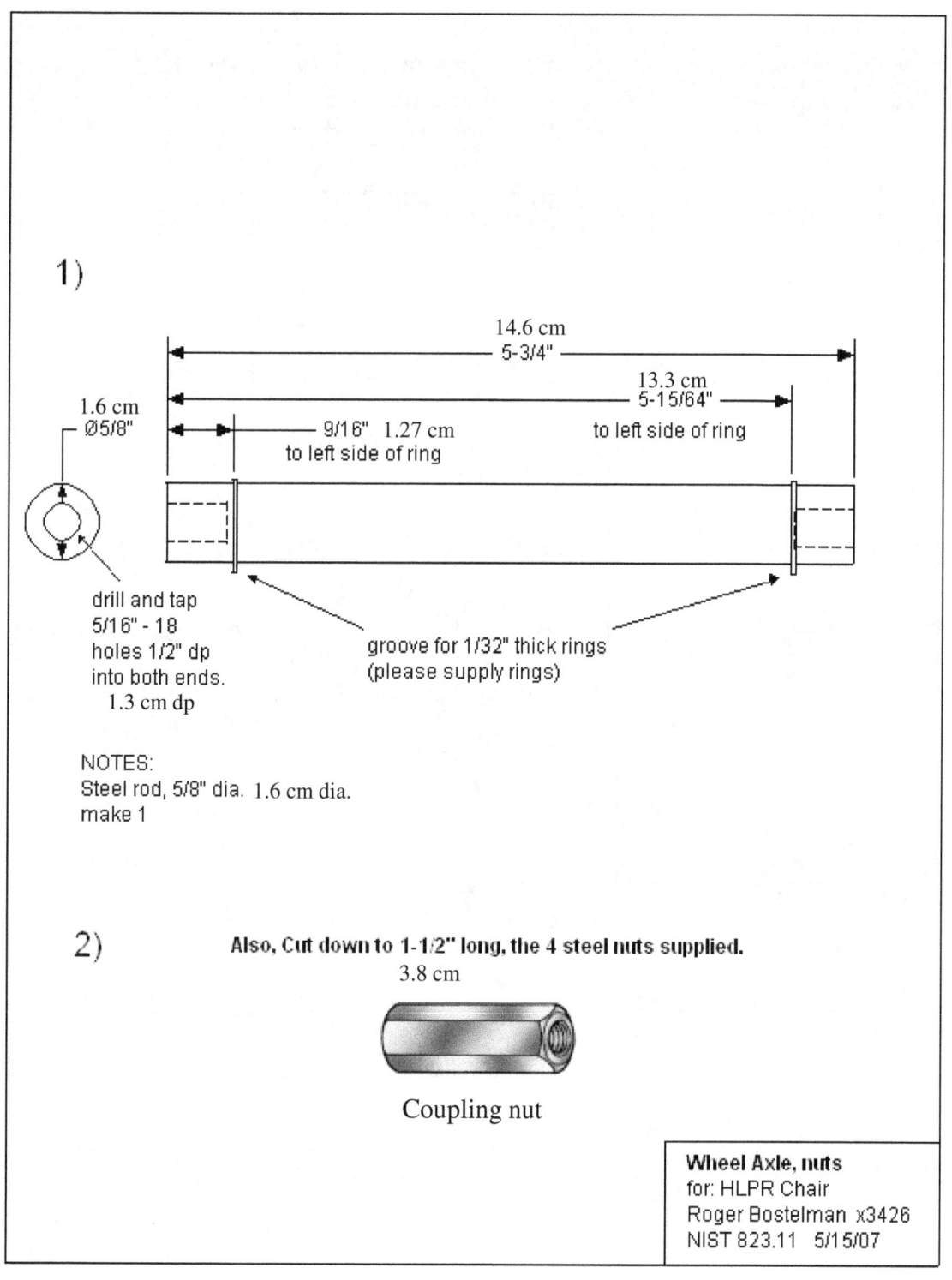

Wheel Axle, nuts
for: HLPR Chair
Roger Bostelman x3426
NIST 823.11 5/15/07

Design of the HLPR Chair

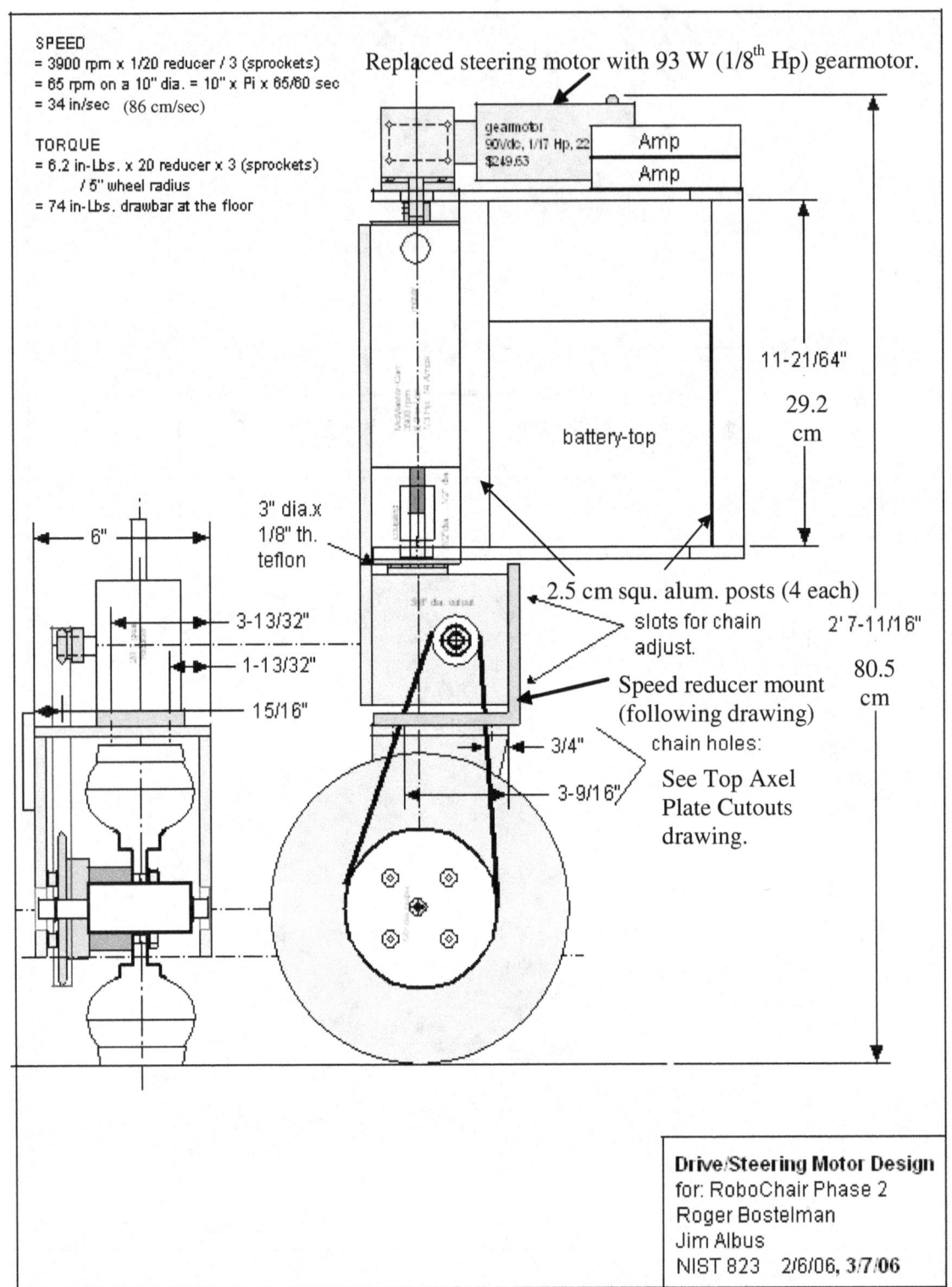

Design of the HLPR Chair

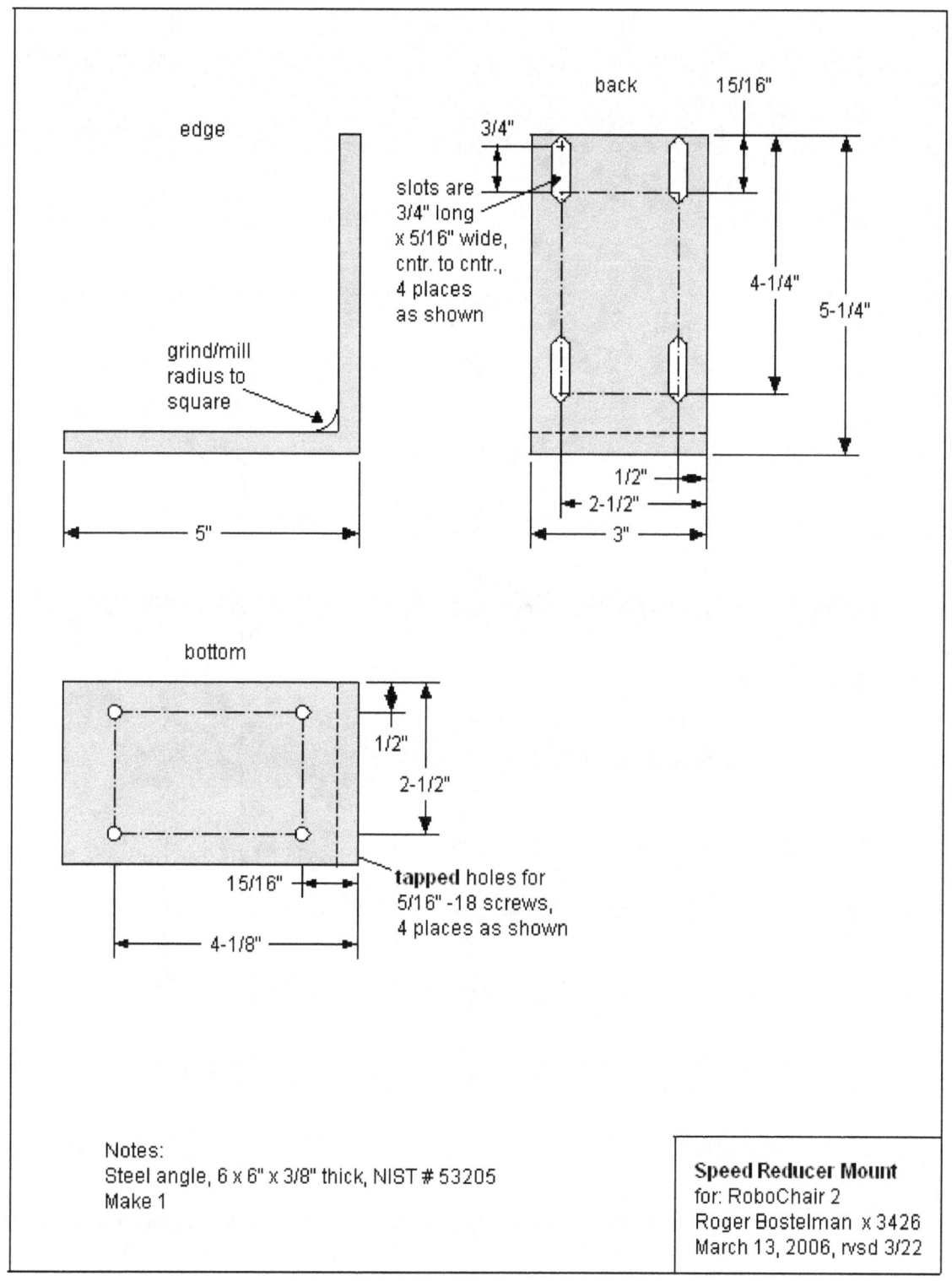

Design of the HLPR Chair

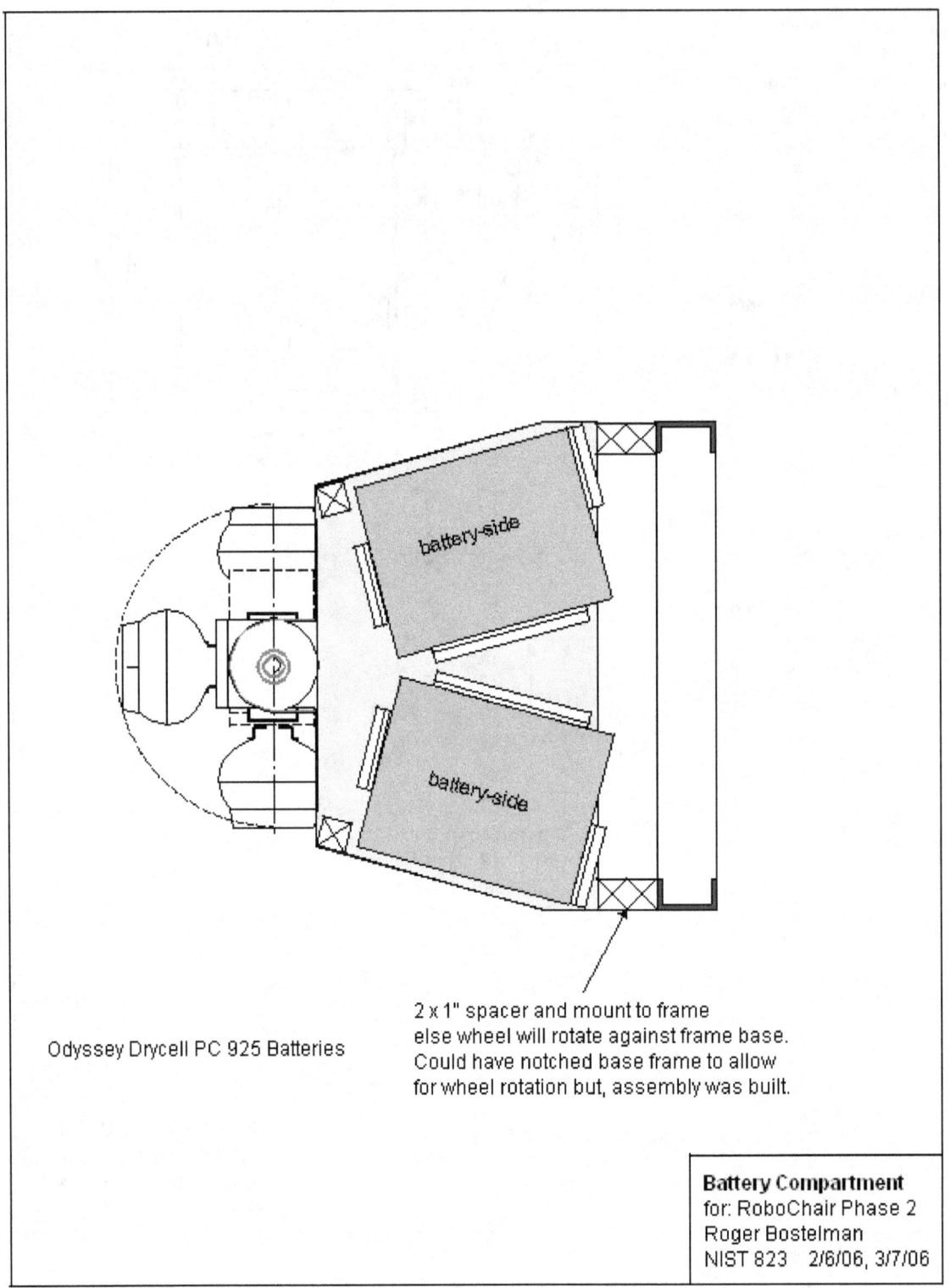

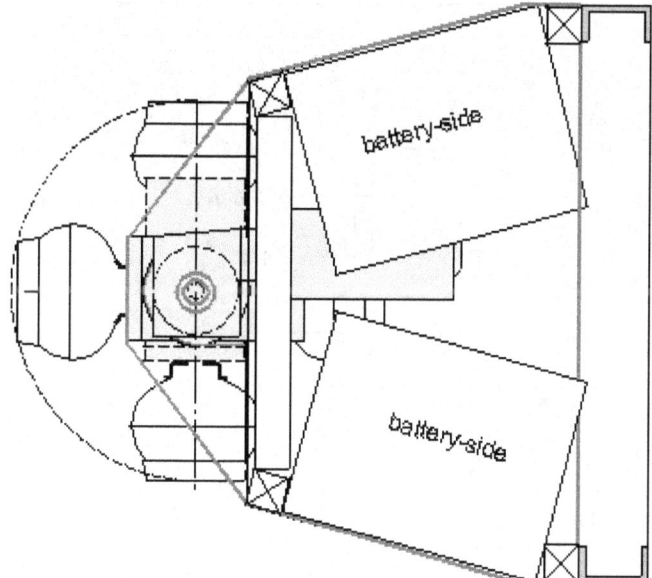

Top view of battery compartment.

Design of the HLPR Chair

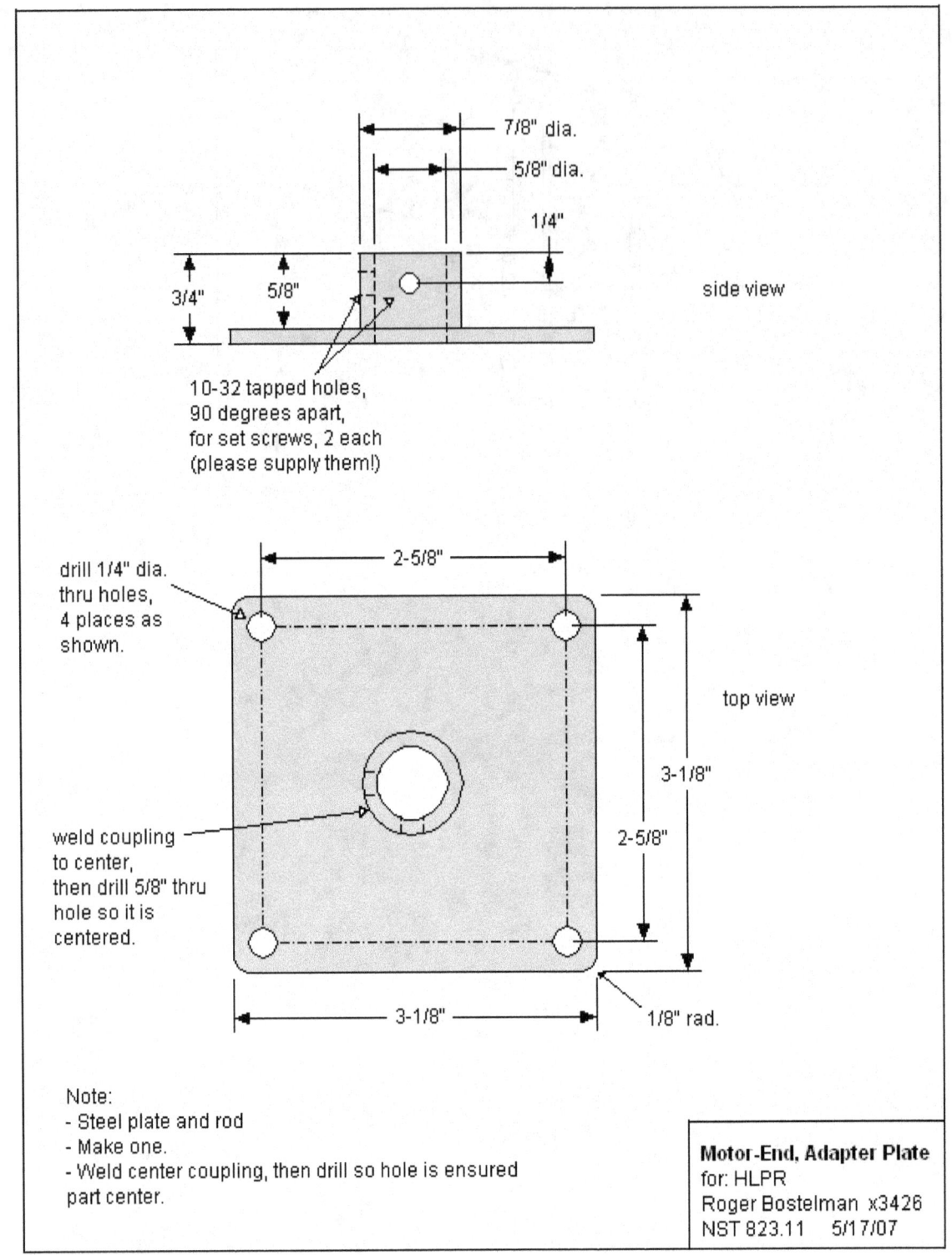

Design of the HLPR Chair

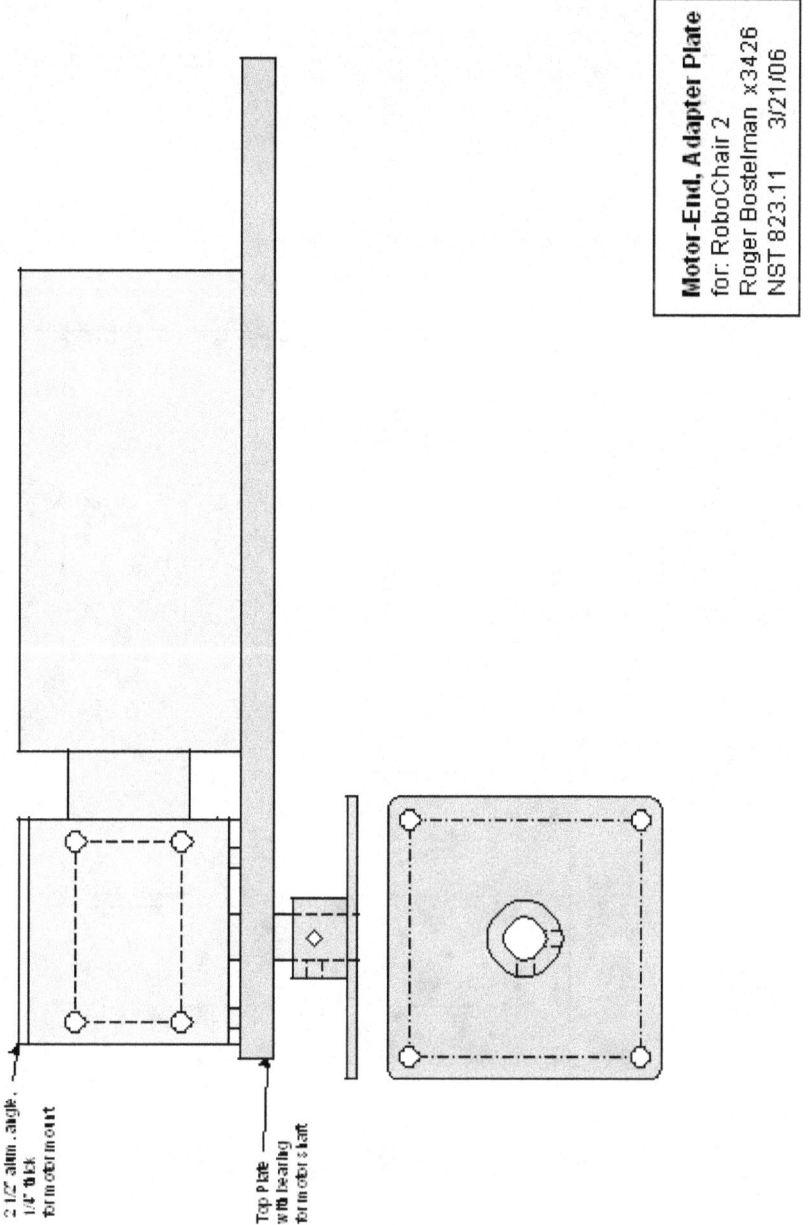

Design of the HLPR Chair

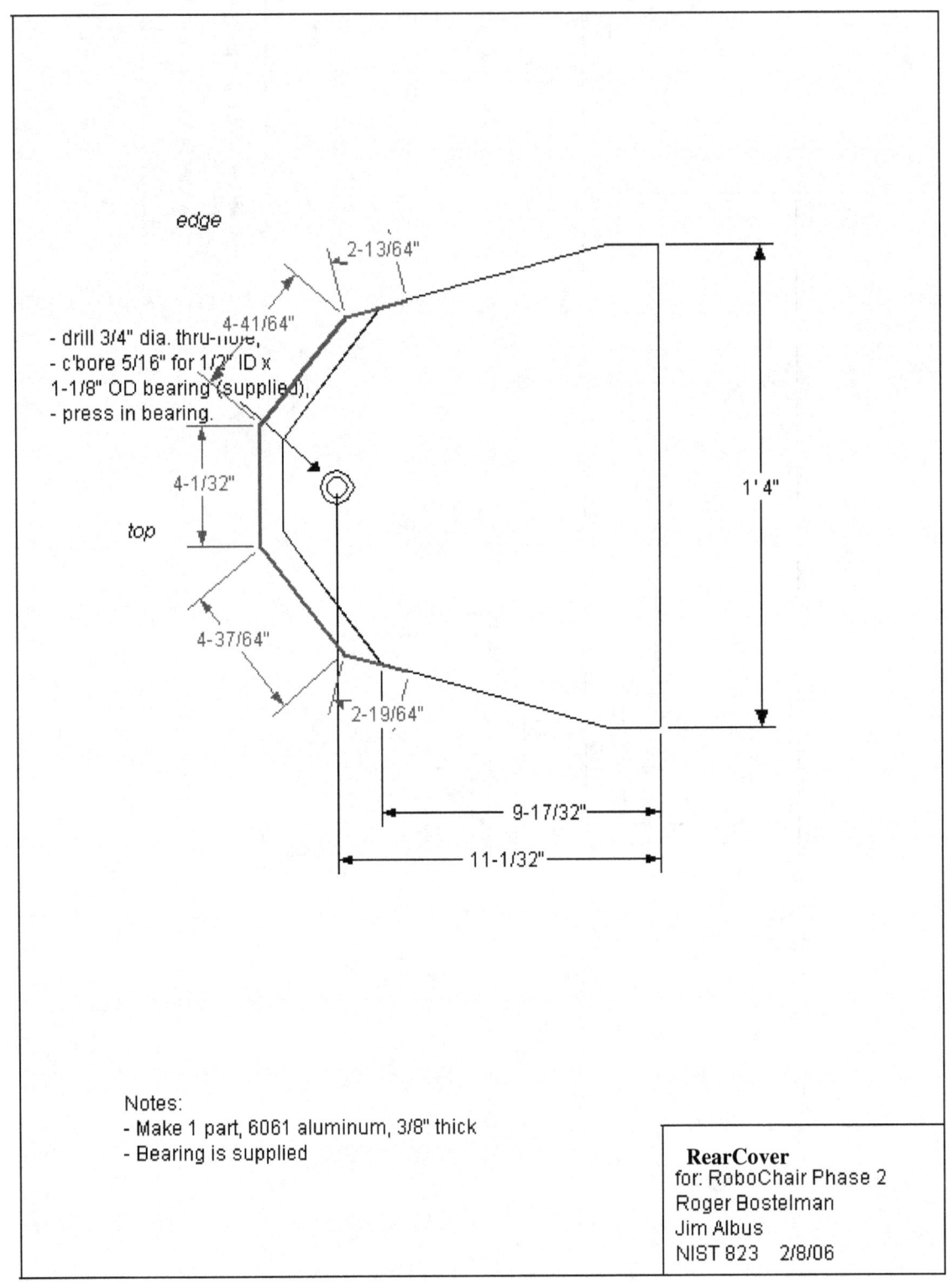

Design of the HLPR Chair

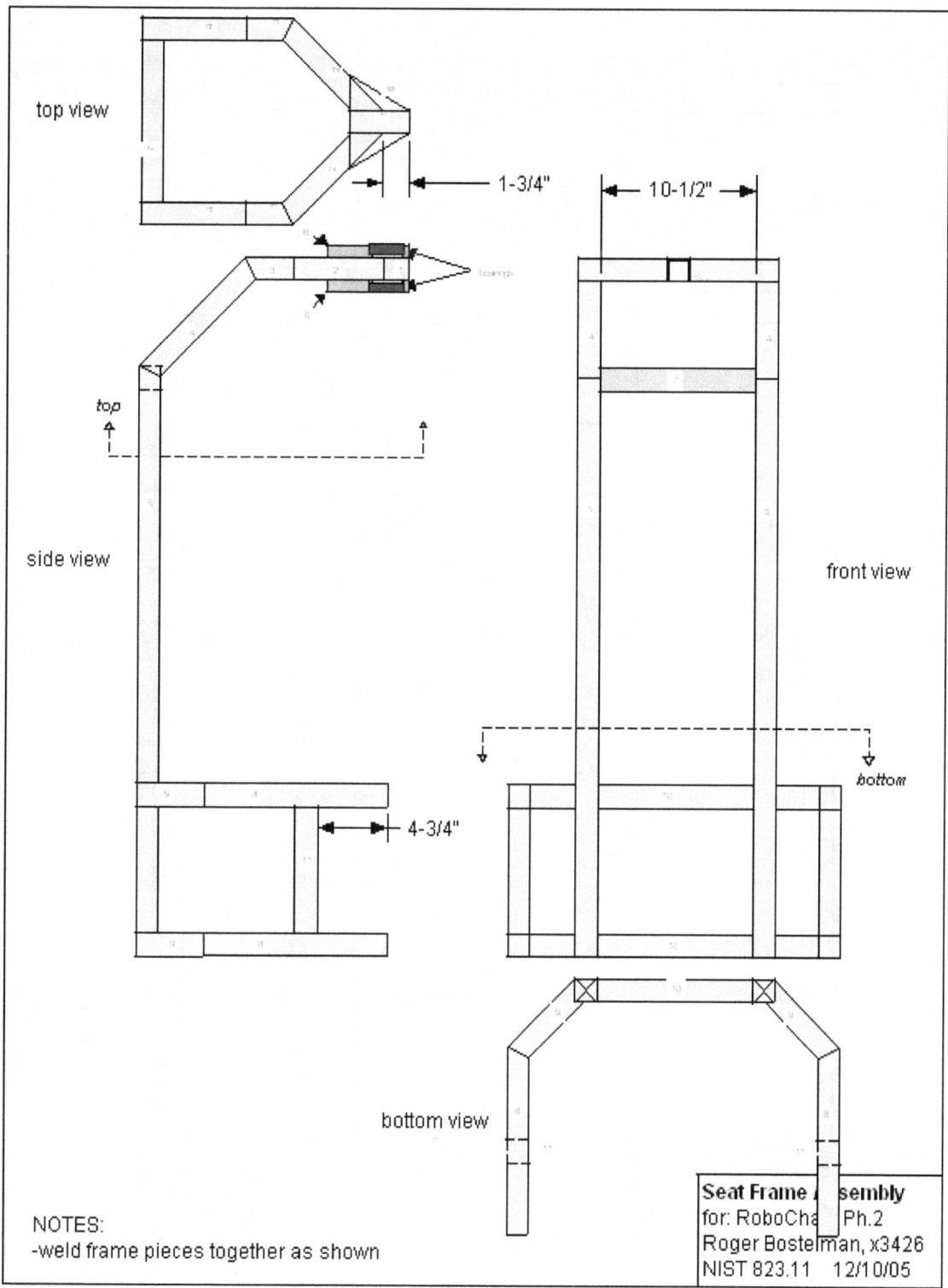

See Modifications to Existing Design section for new version of the seat frame assembly.

Design of the HLPR Chair

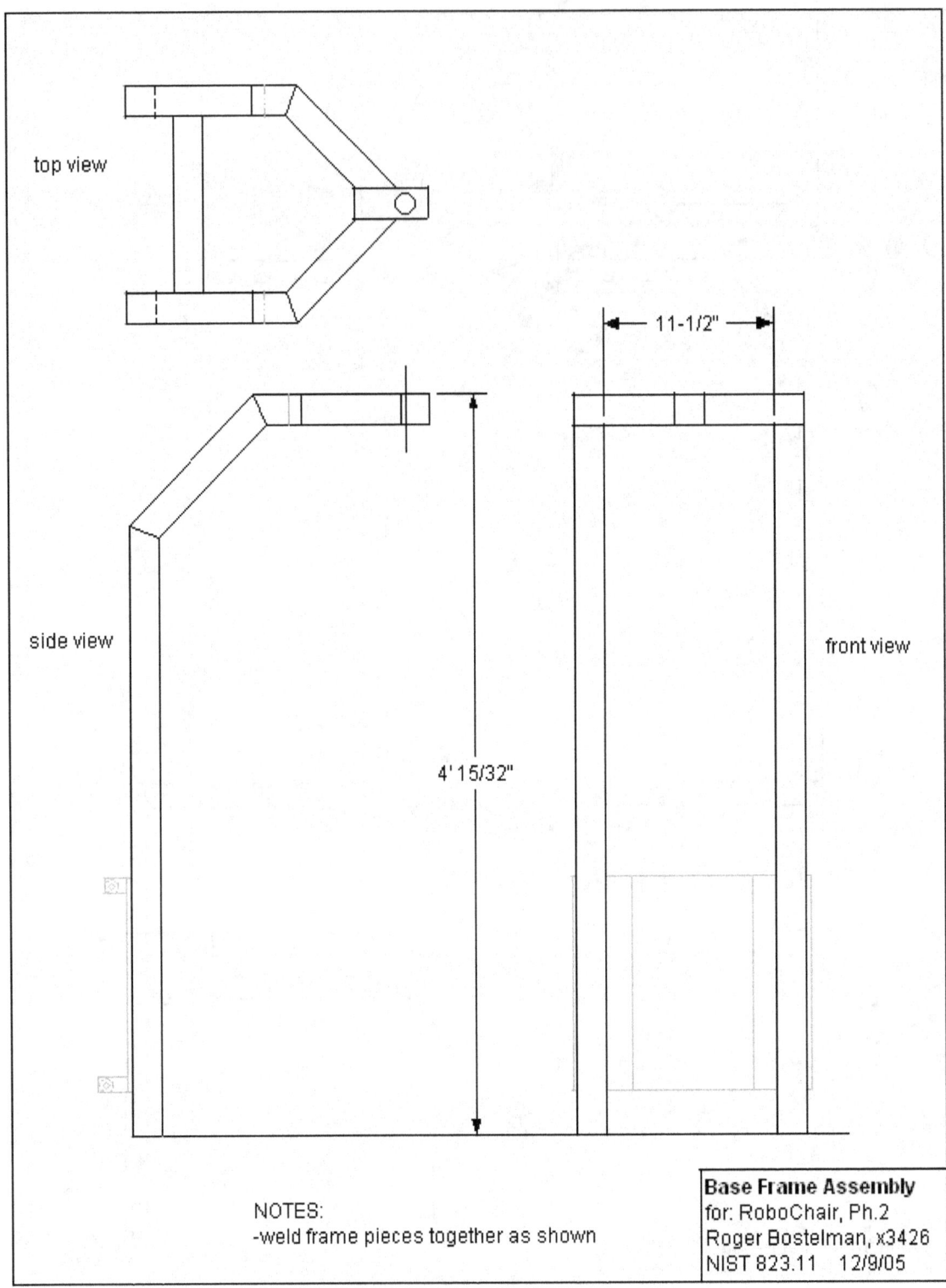

See Modifications to Existing Design section for new version of the base frame assembly.

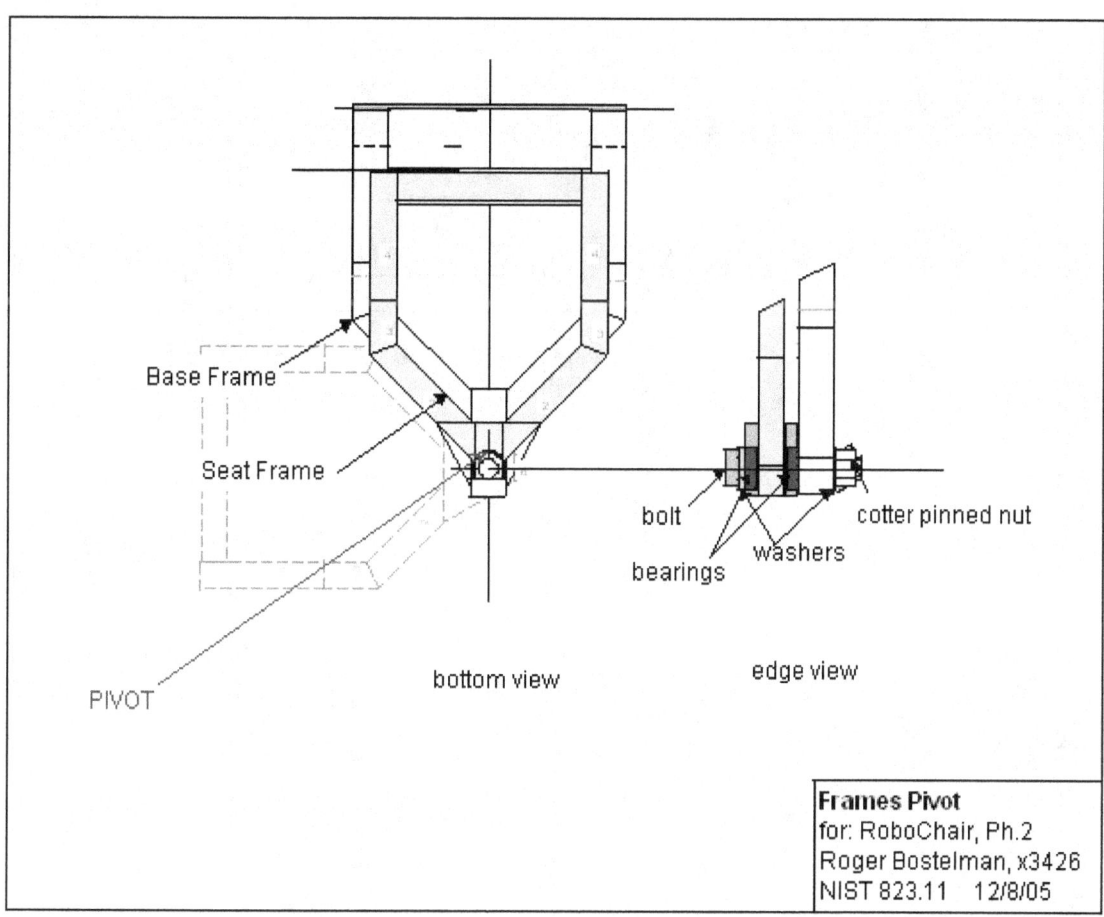

See Modifications to Existing Design section for new version of the frames pivot.

Design of the HLPR Chair

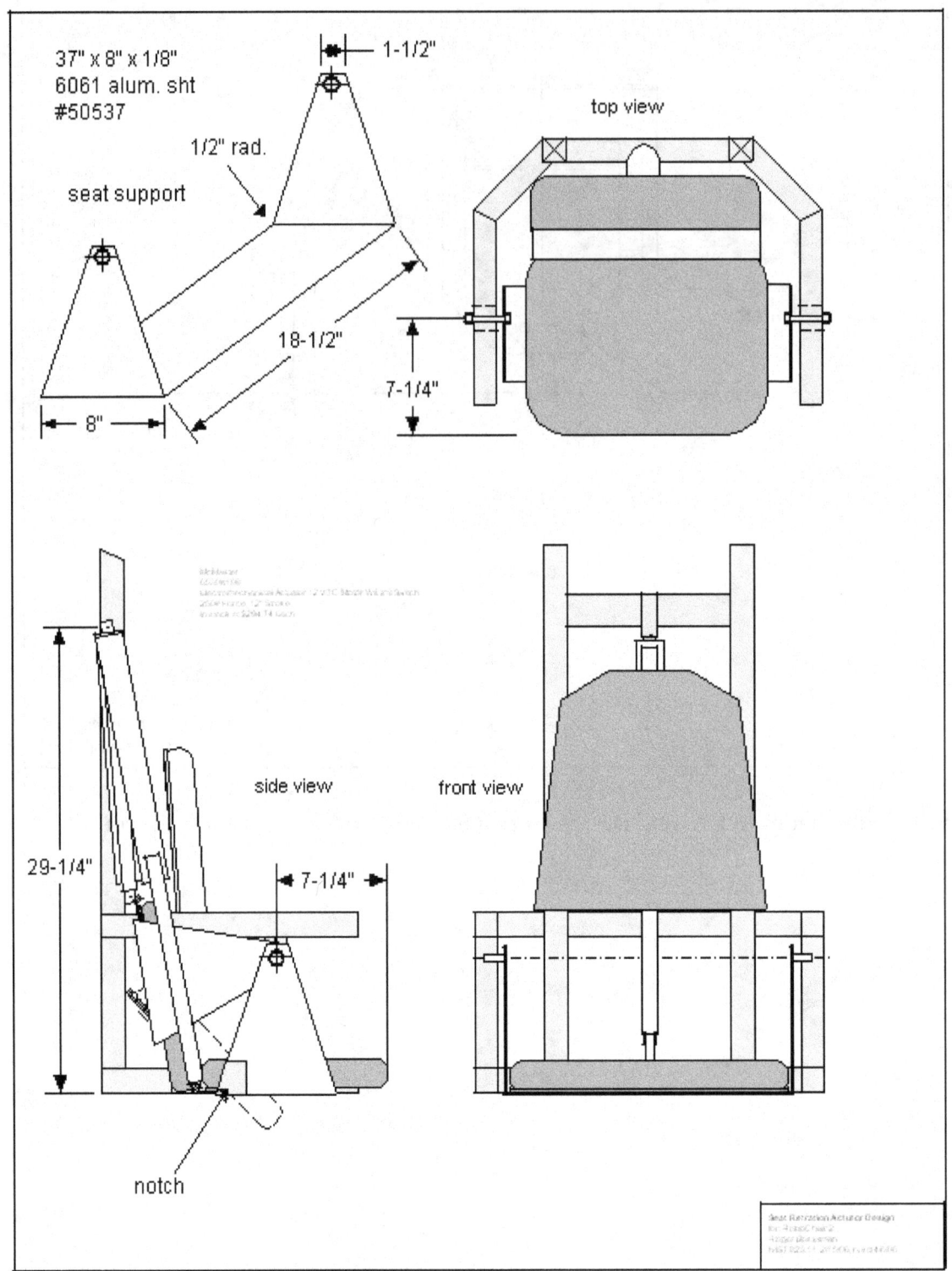

Design of the HLPR Chair

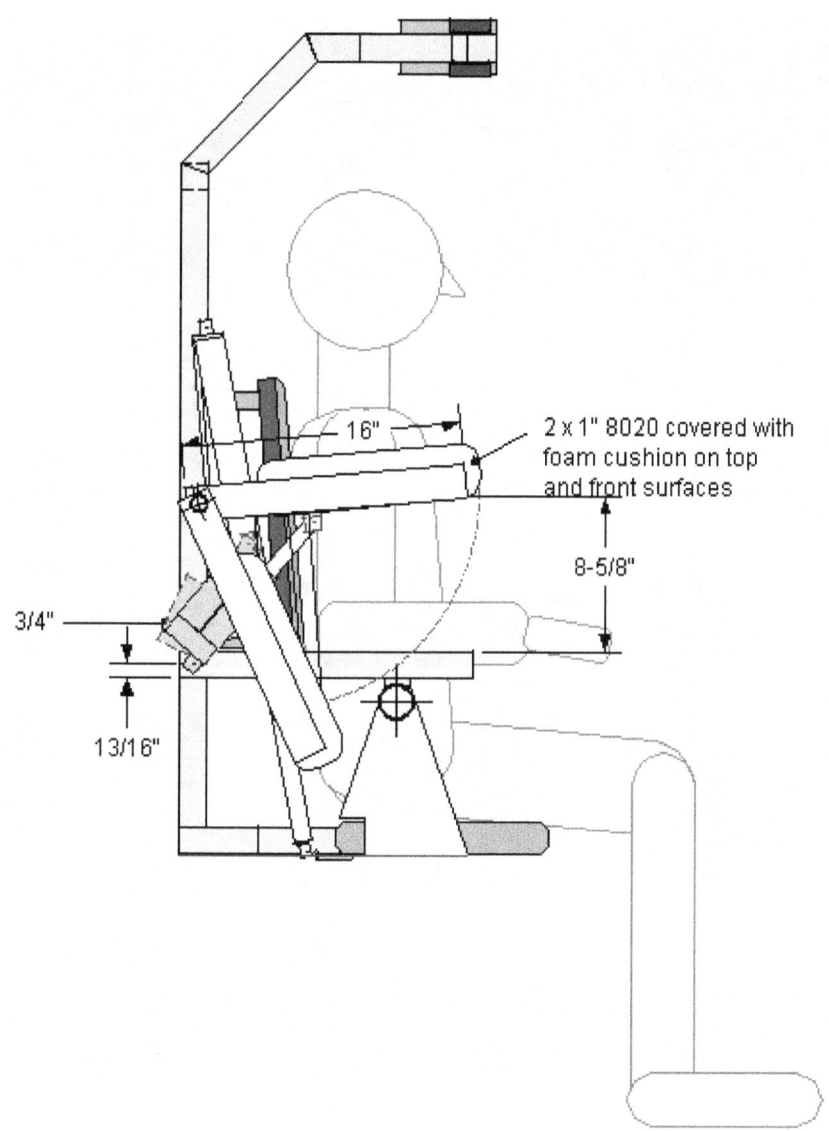

Torso Lift Design
for: RoboChair 2
Roger Bostelman
NIST 823.11 4/24/06

Design of the HLPR Chair

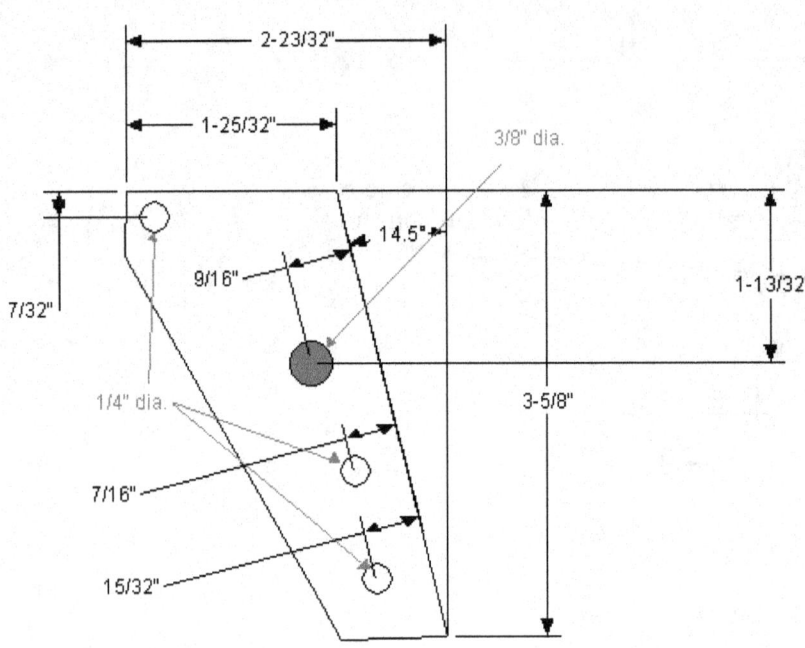

Design of the HLPR Chair

actuator support bracket

Make 1

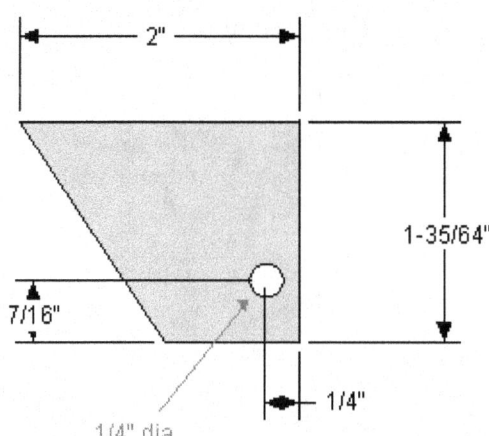

Design of the HLPR Chair

foot rest
Make 2

Design of the HLPR Chair

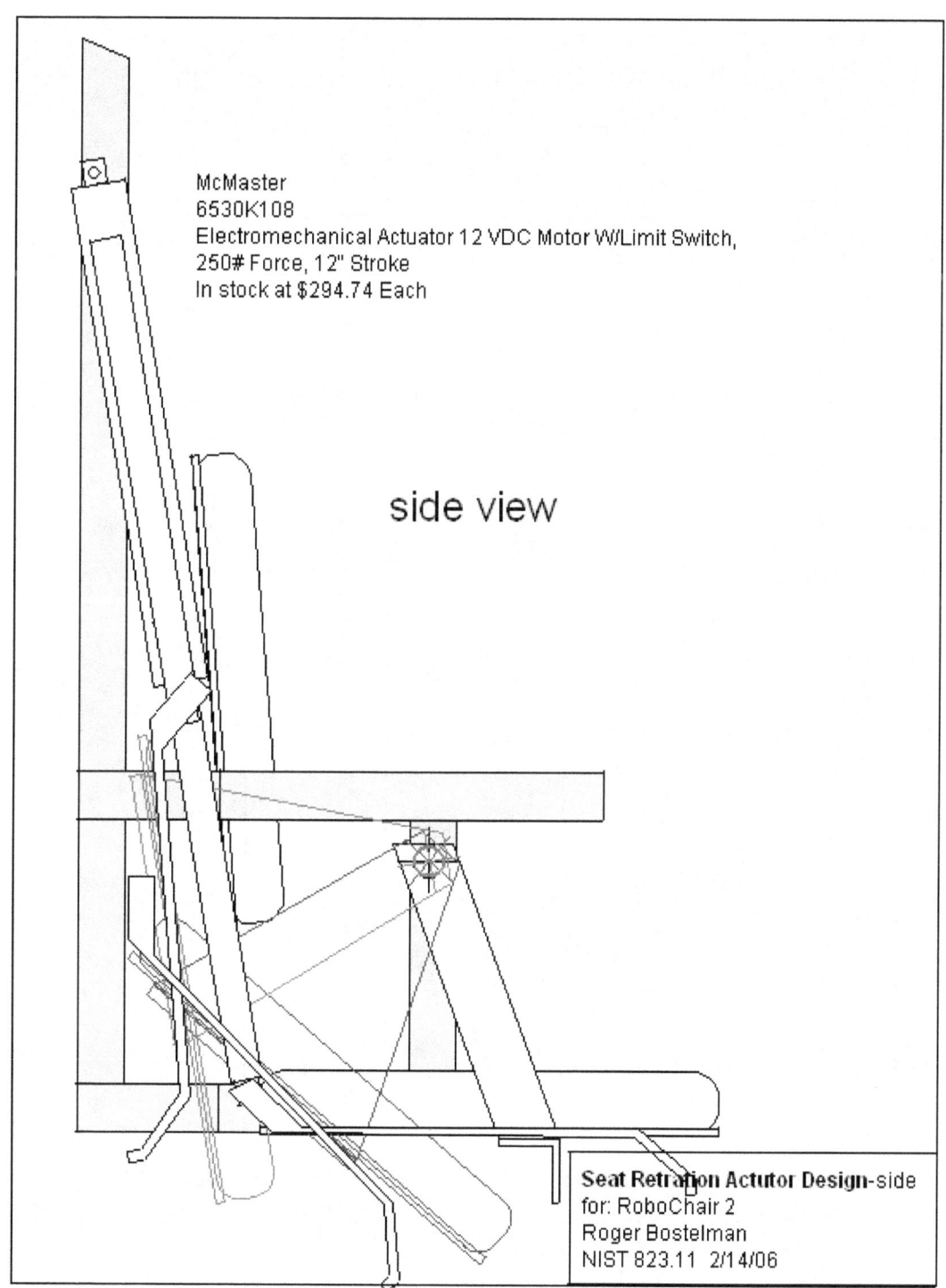

Design of the HLPR Chair

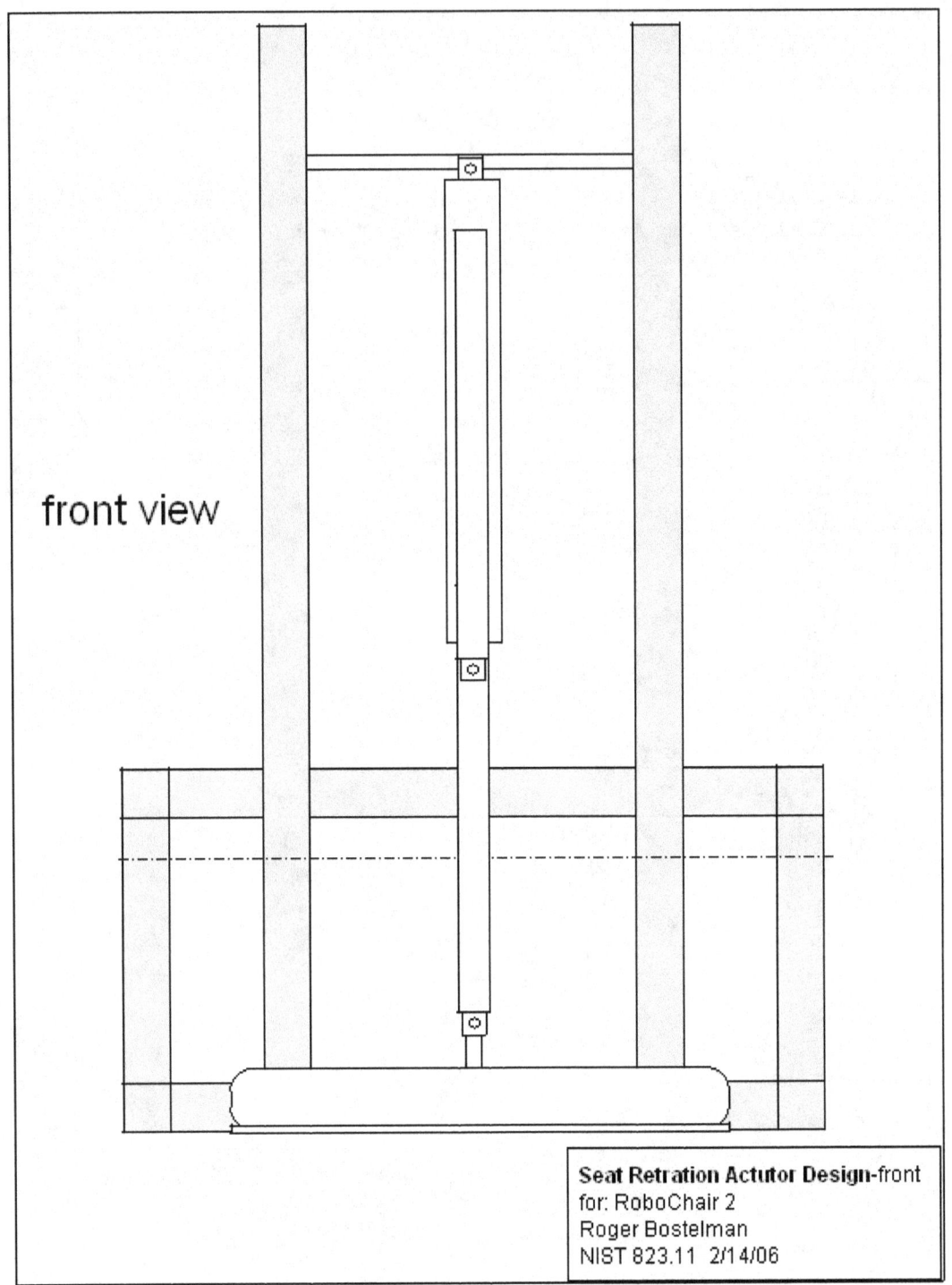

Design of the HLPR Chair

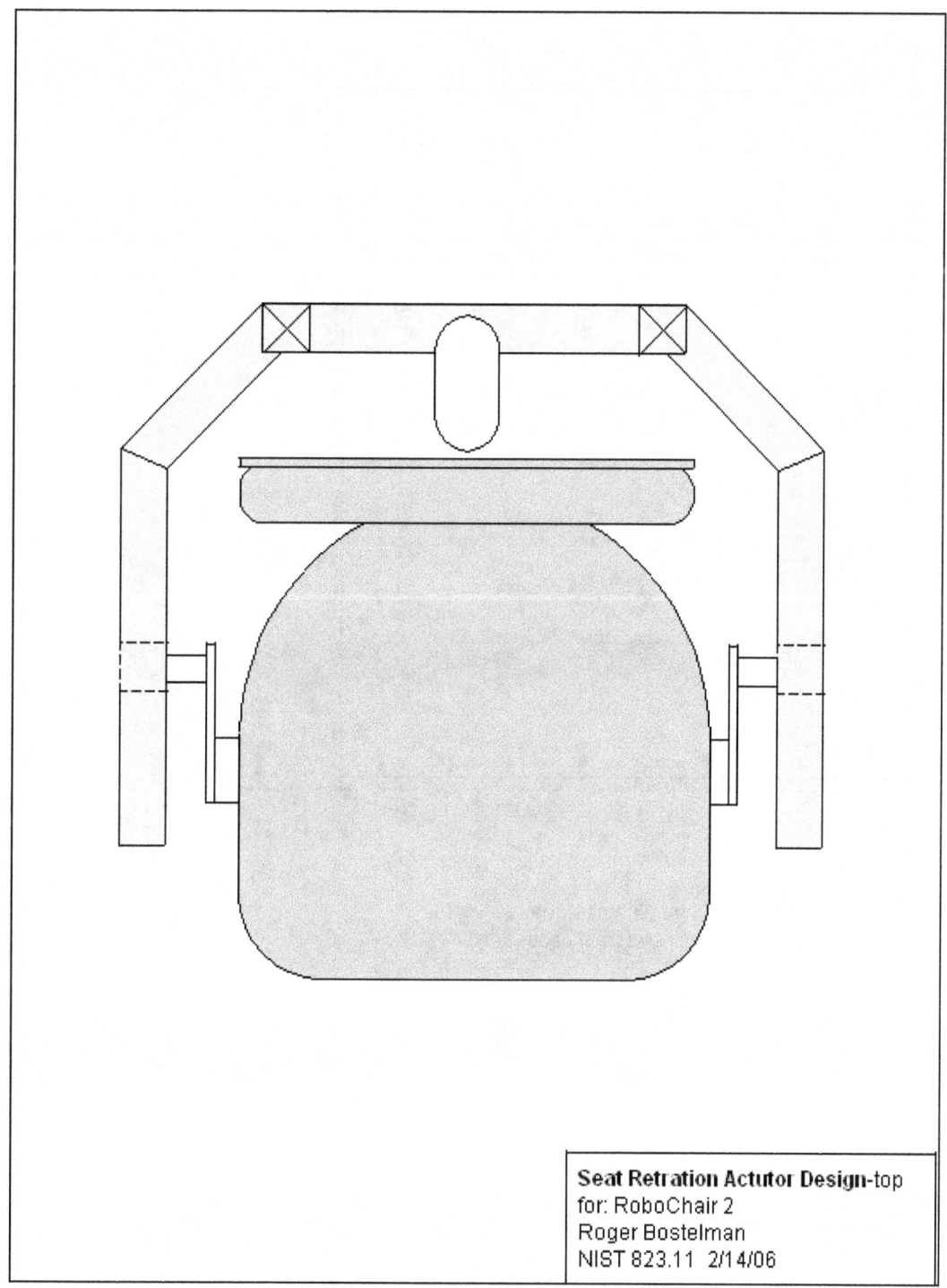

Seat Retration Actutor Design-top
for: RoboChair 2
Roger Bostelman
NIST 823.11 2/14/06

Design of the HLPR Chair

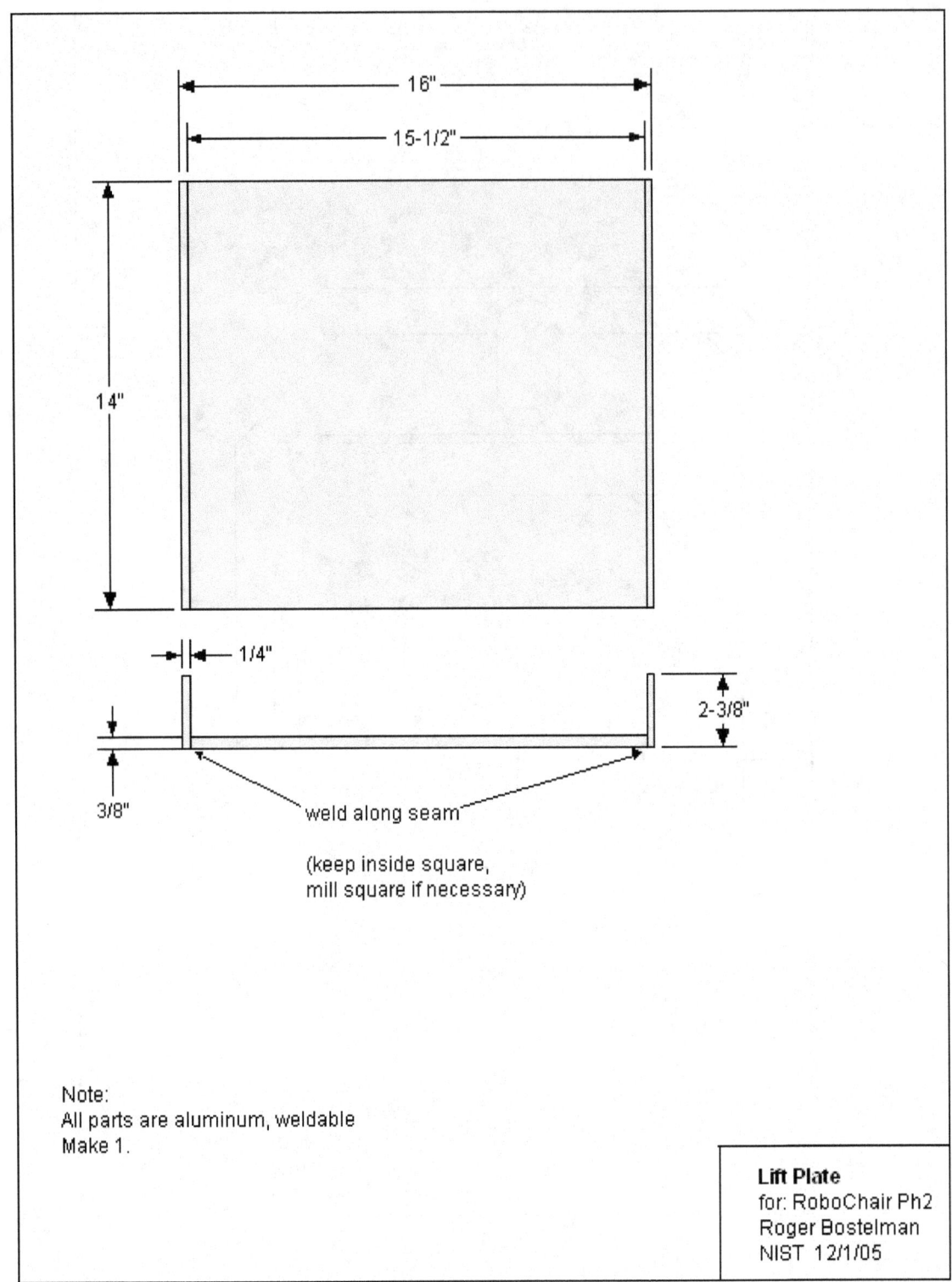

Note:
All parts are aluminum, weldable
Make 1.

Lift Plate
for: RoboChair Ph2
Roger Bostelman
NIST 12/1/05

Design of the HLPR Chair

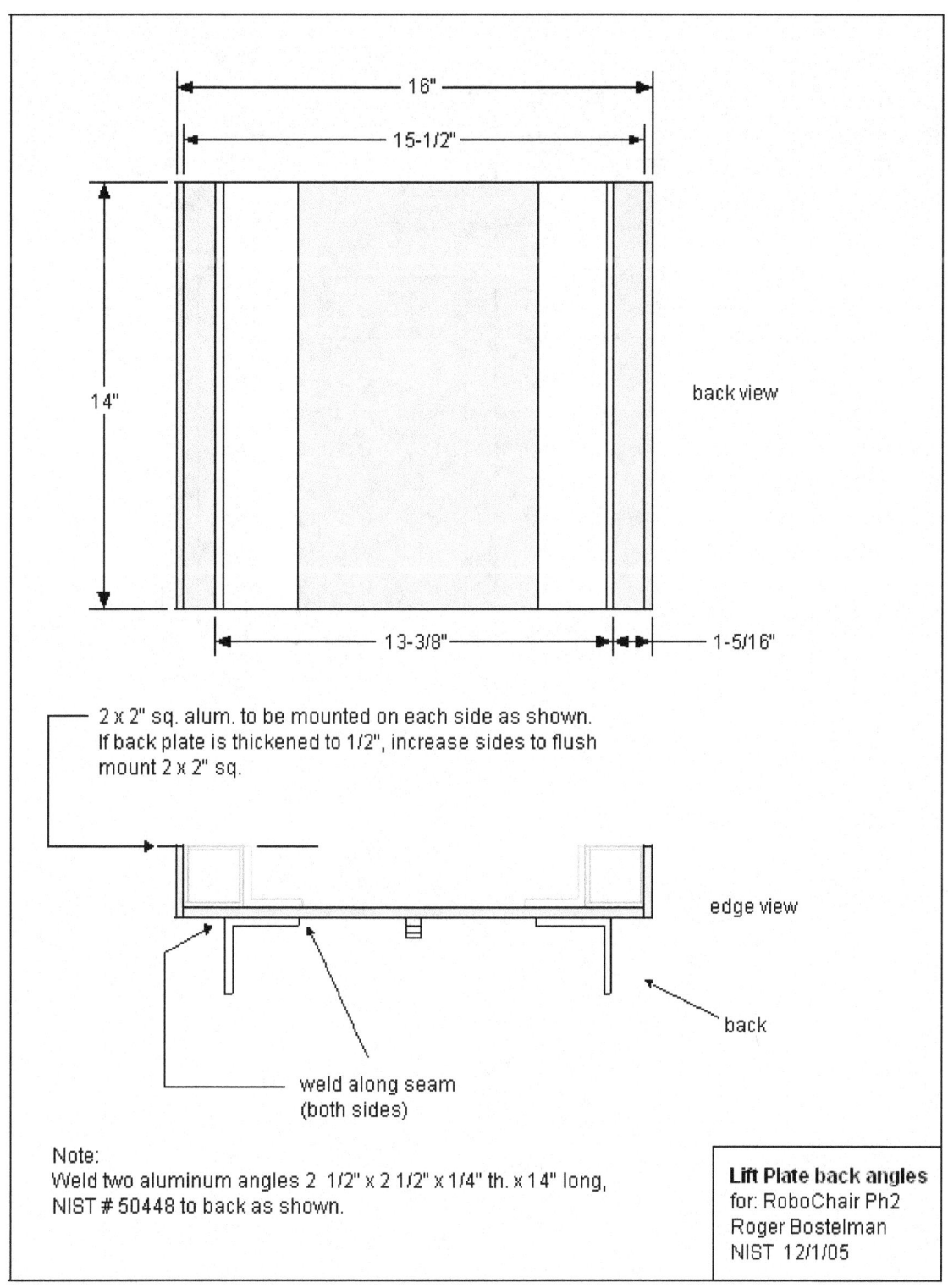

Design of the HLPR Chair

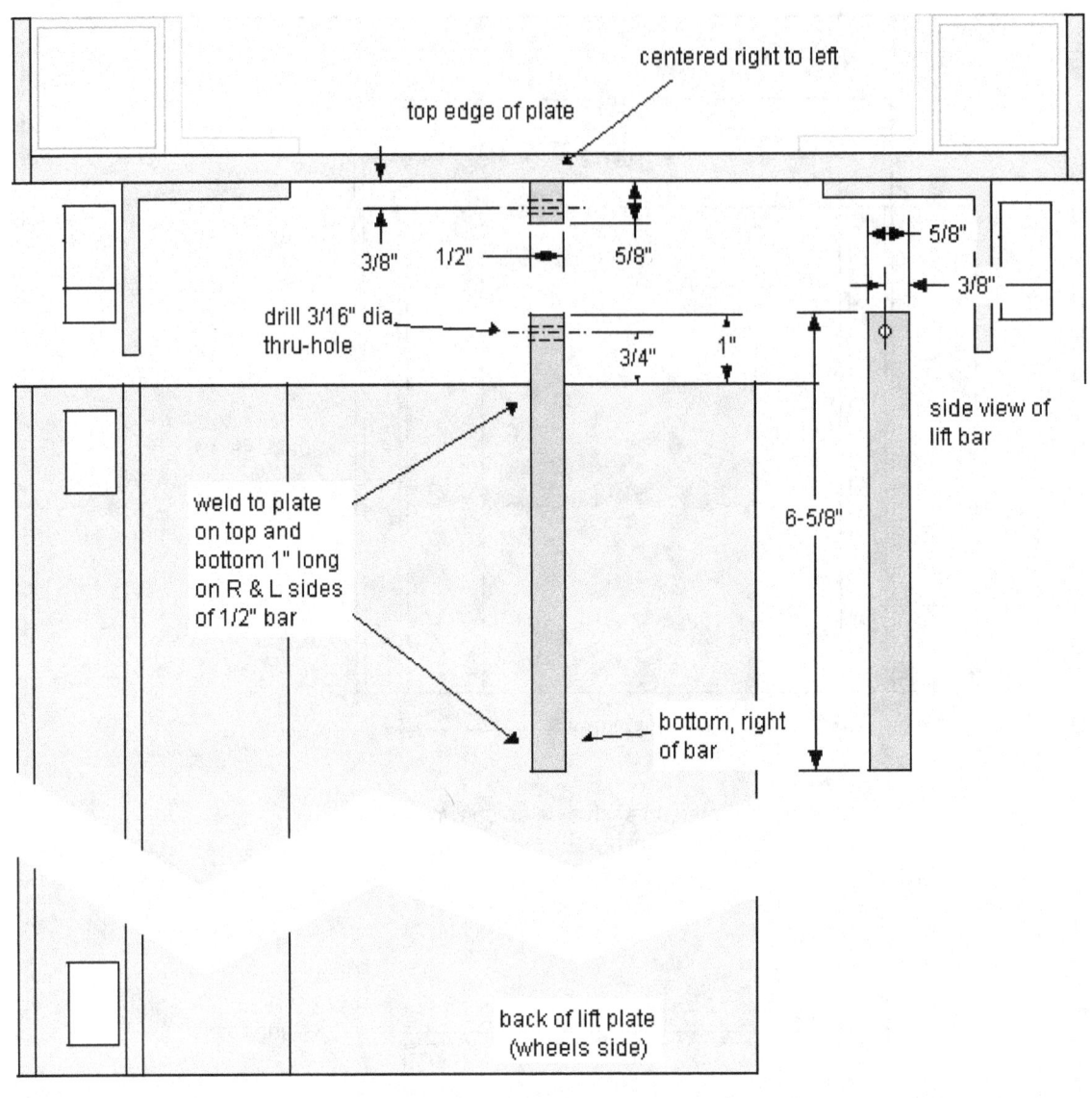

Notes:
Make one part, weld it to lift plate (supplied)
Aluminum, 6061 bar stock.

Lift Bar
for: RoboChair 2
Roger Bostelman 2/10/06
NIST 823.11

Design of the HLPR Chair

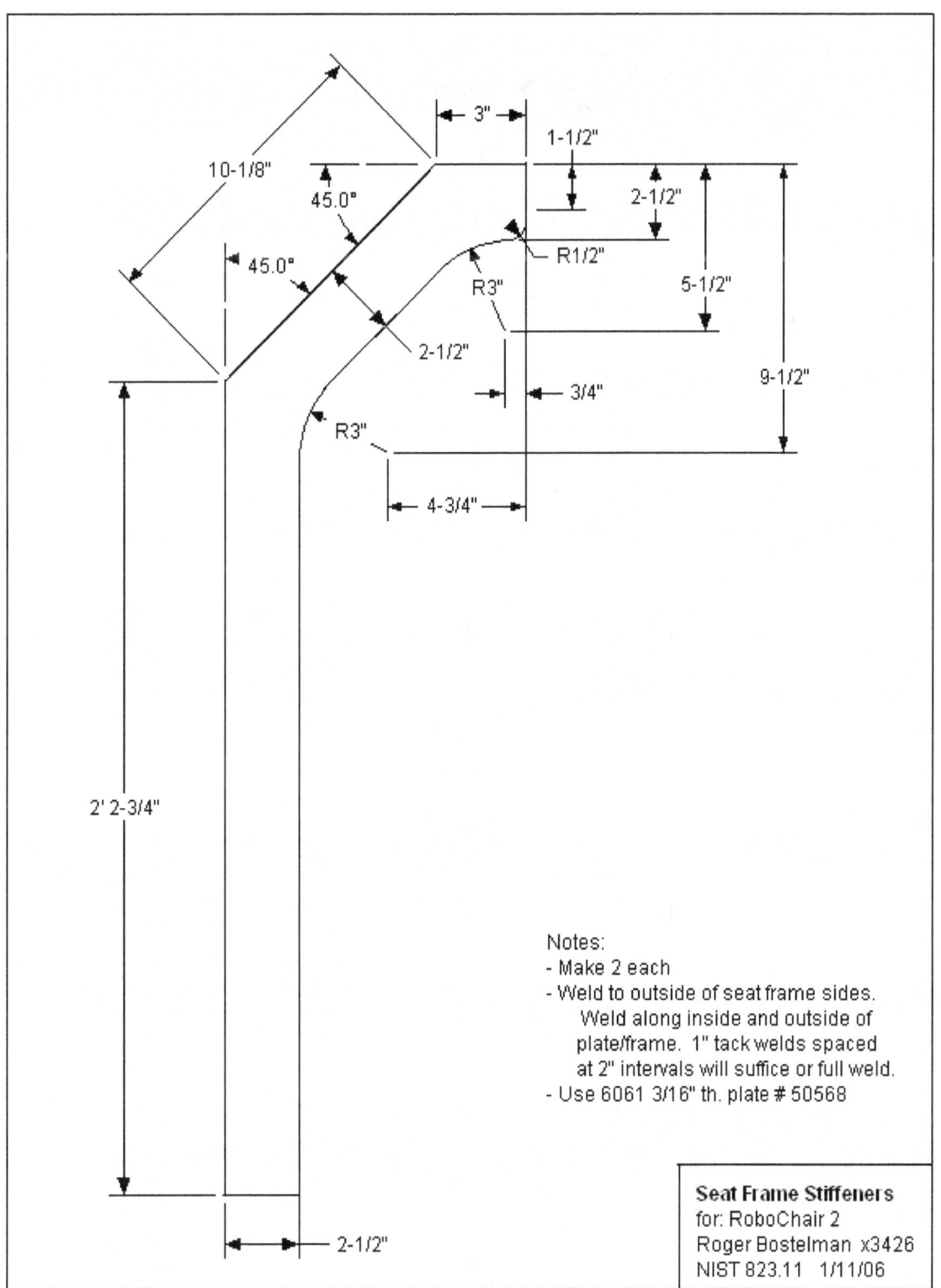

Design of the HLPR Chair

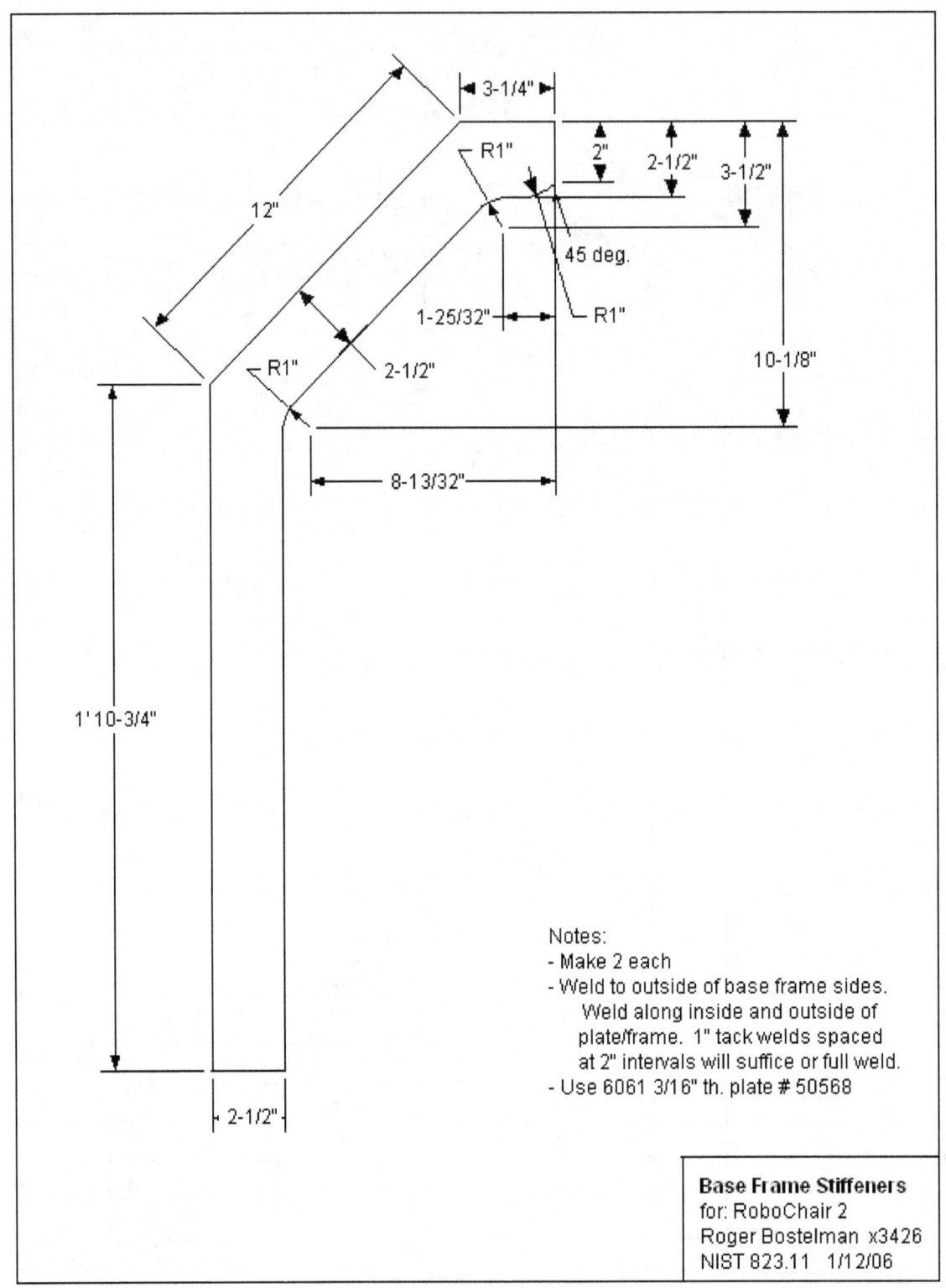

Design of the HLPR Chair

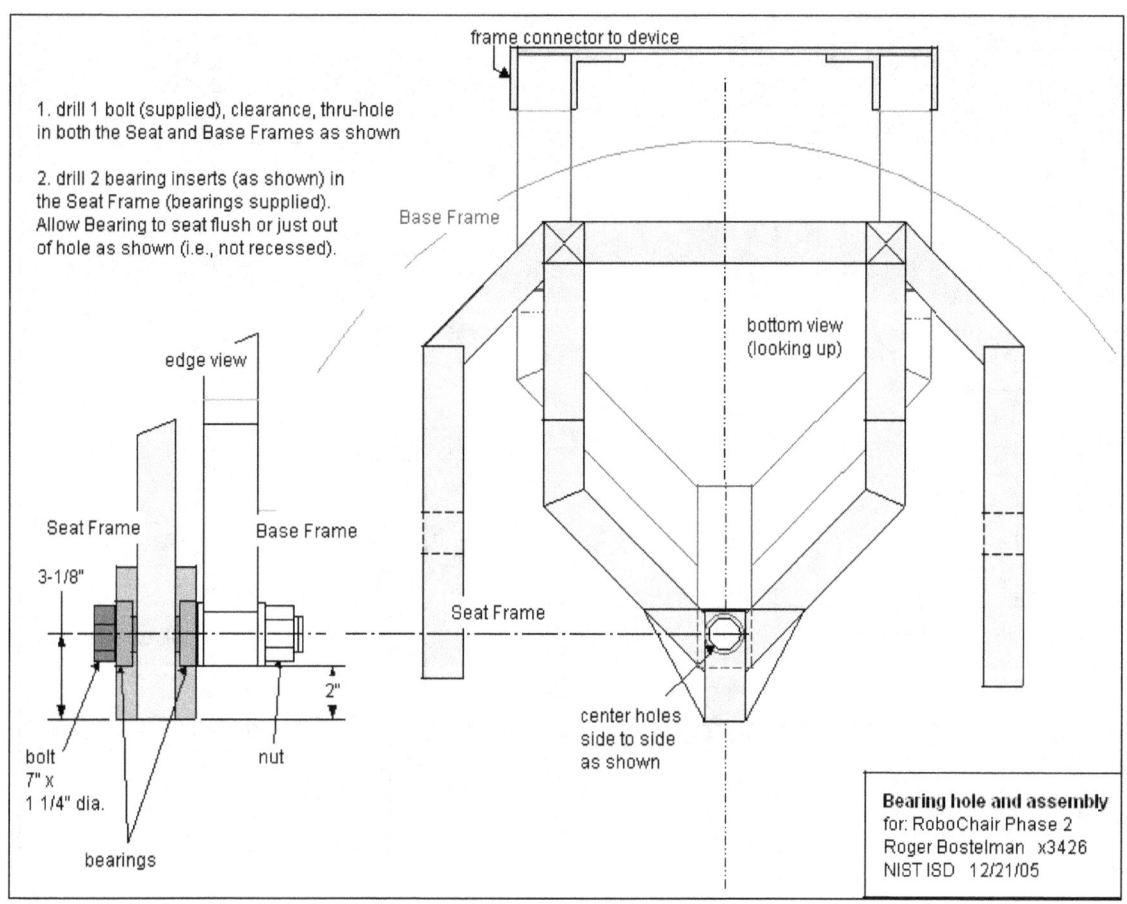

Design of the HLPR Chair

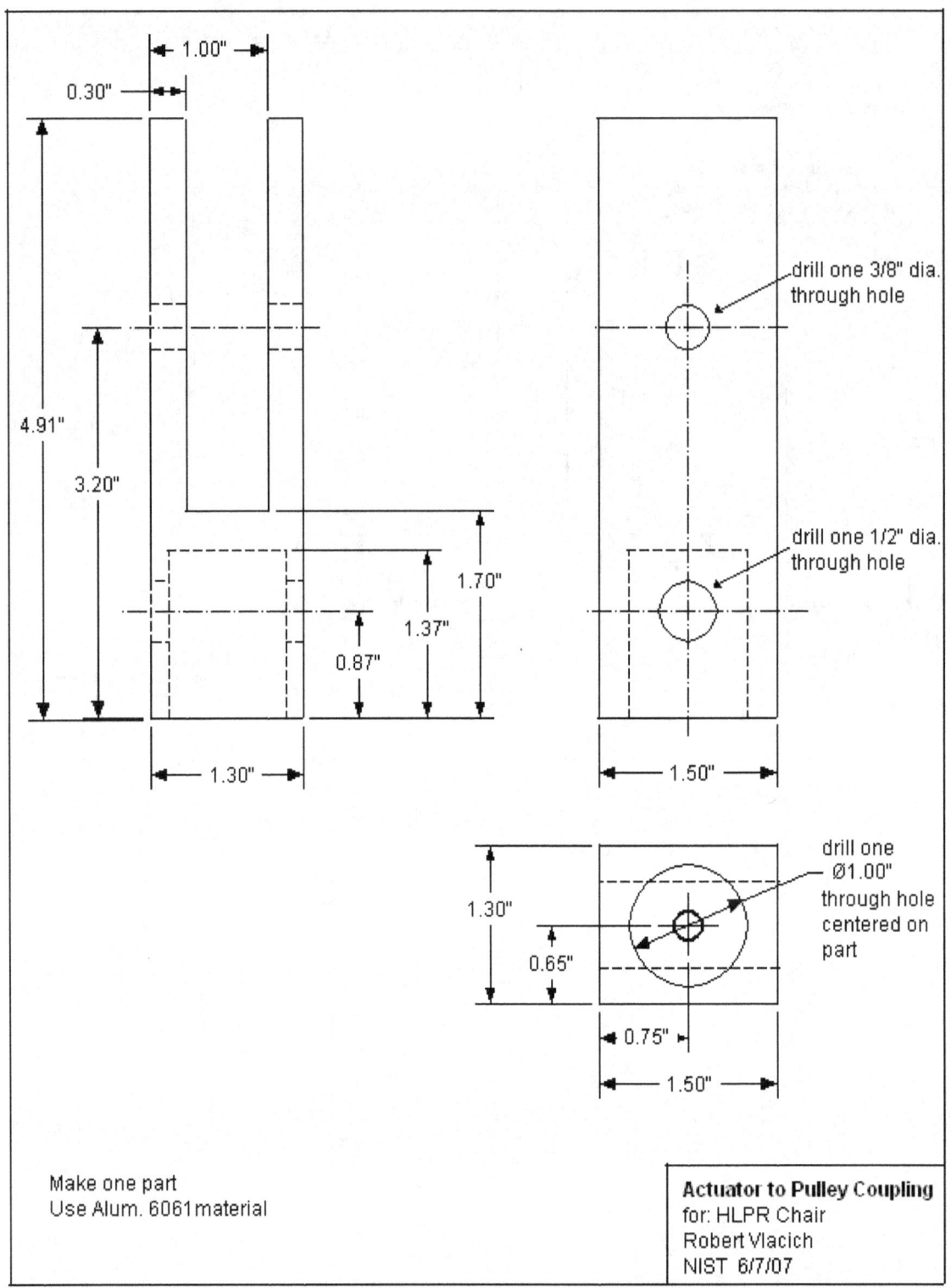

Modifications to Original Designs

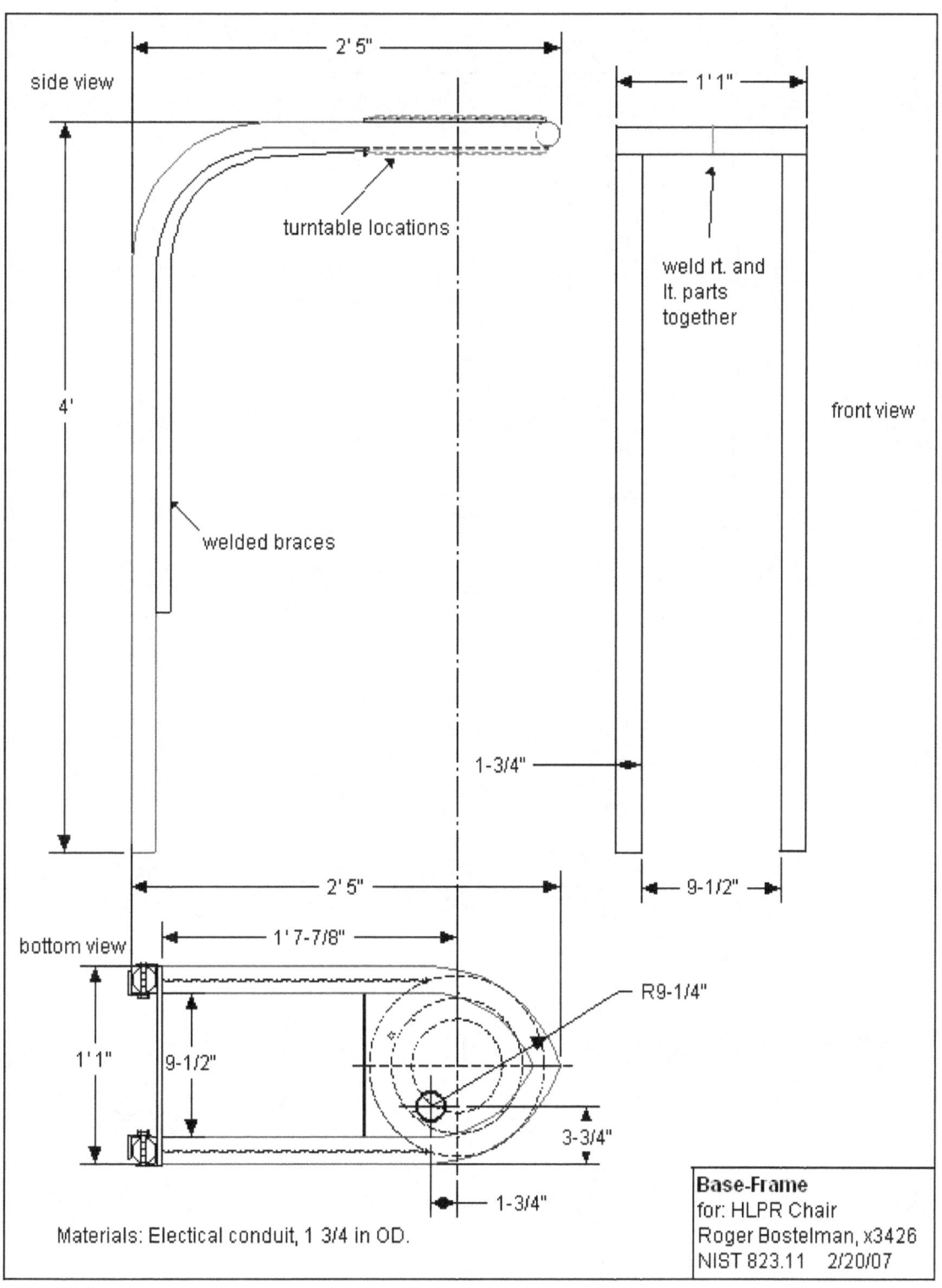

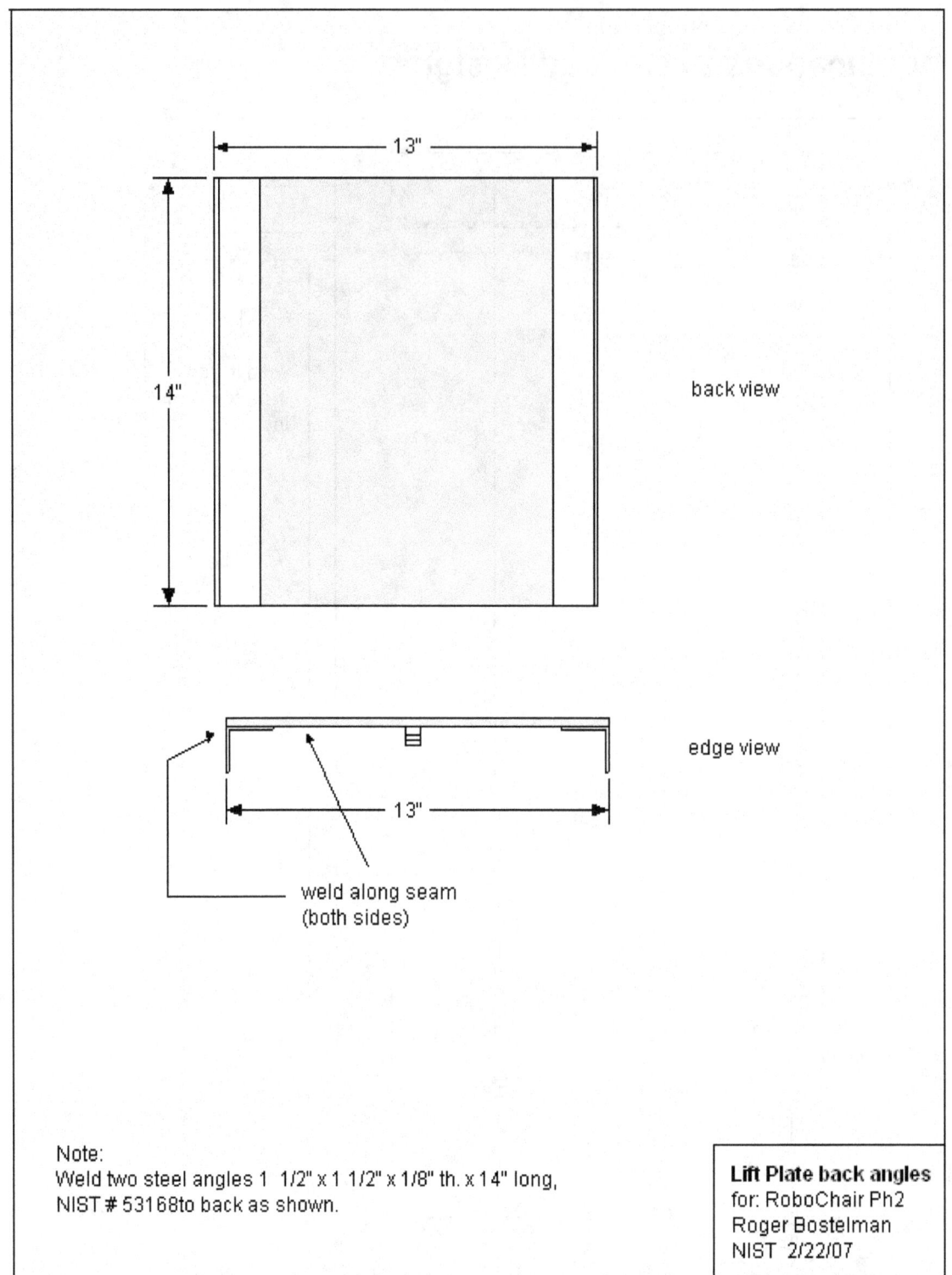

Design of the HLPR Chair

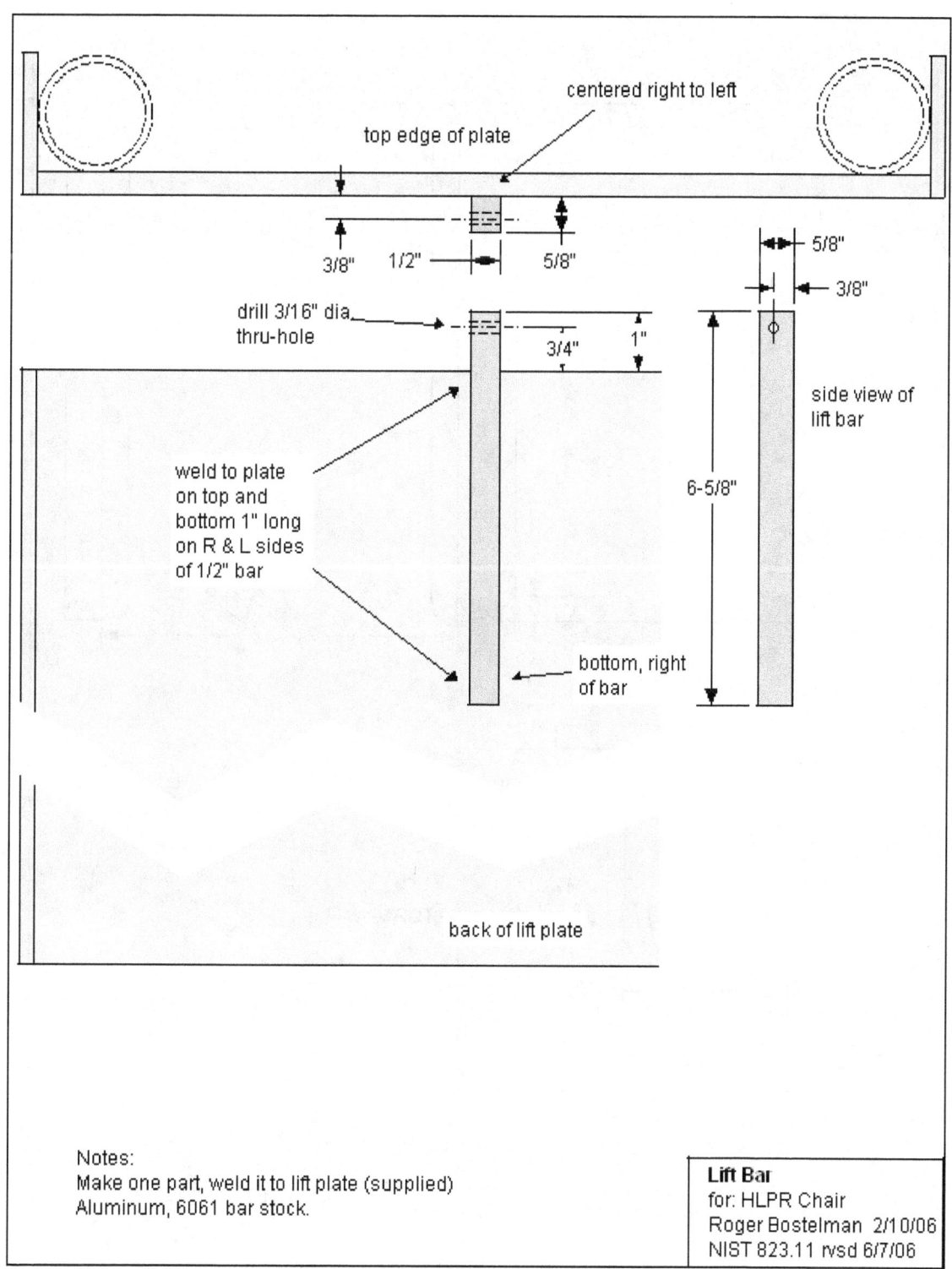

Design of the HLPR Chair

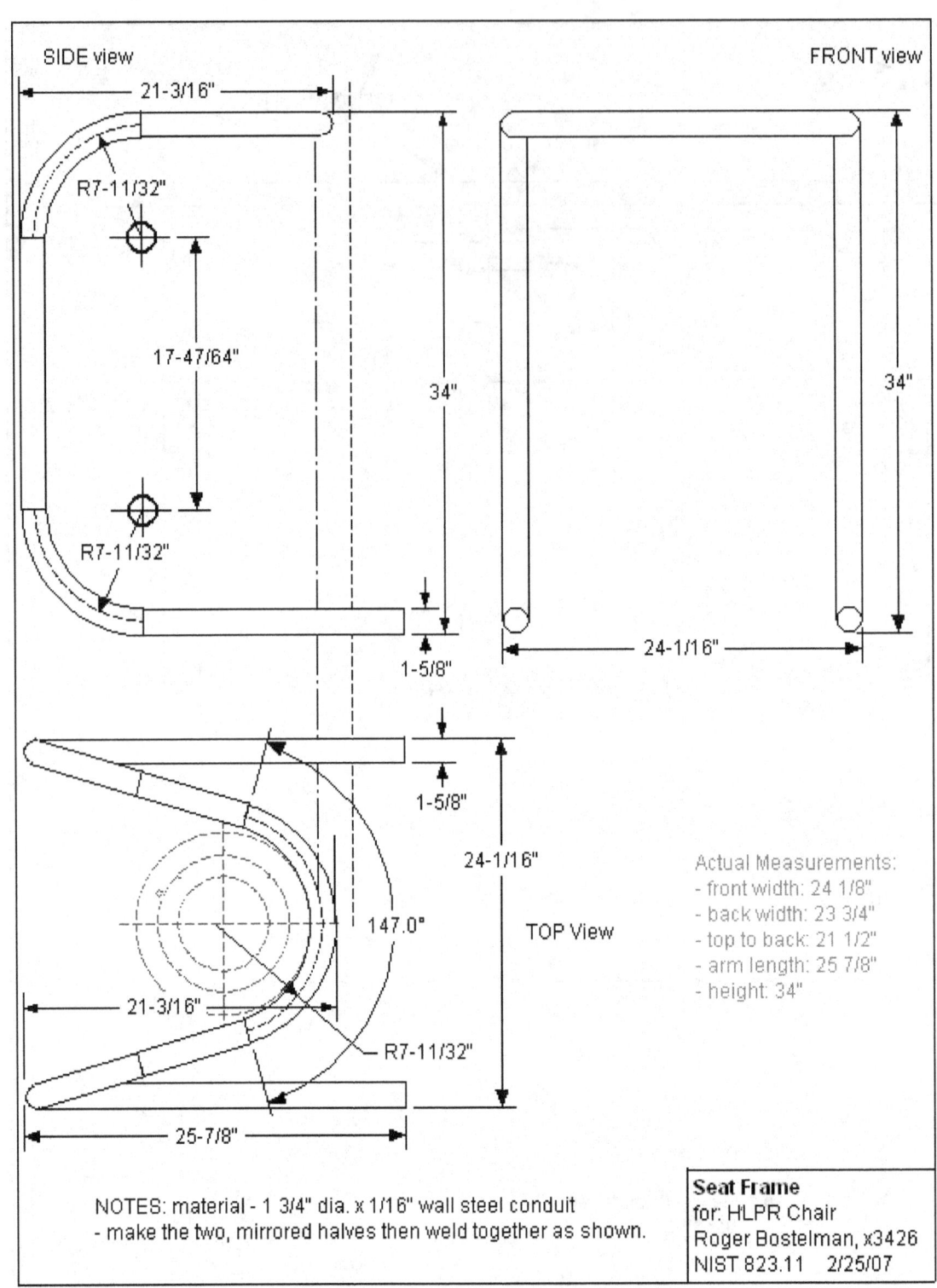

Design of the HLPR Chair

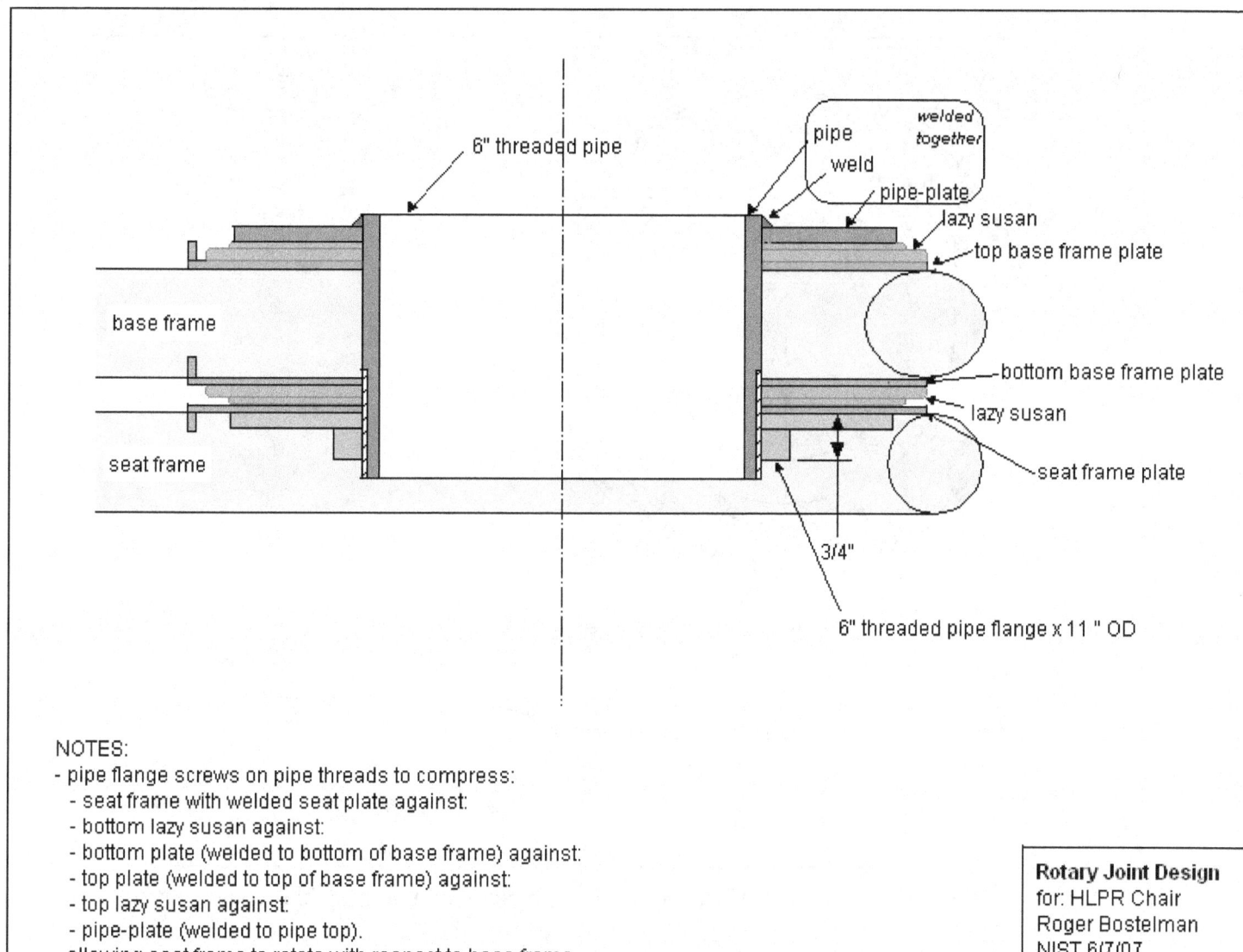

NOTES:
- pipe flange screws on pipe threads to compress:
  - seat frame with welded seat plate against:
  - bottom lazy susan against:
  - bottom plate (welded to bottom of base frame) against:
  - top plate (welded to top of base frame) against:
  - top lazy susan against:
  - pipe-plate (welded to pipe top).
- allowing seat frame to rotate with respect to base frame.

Rotary Joint Design
for: HLPR Chair
Roger Bostelman
NIST 6/7/07

Design of the HLPR Chair

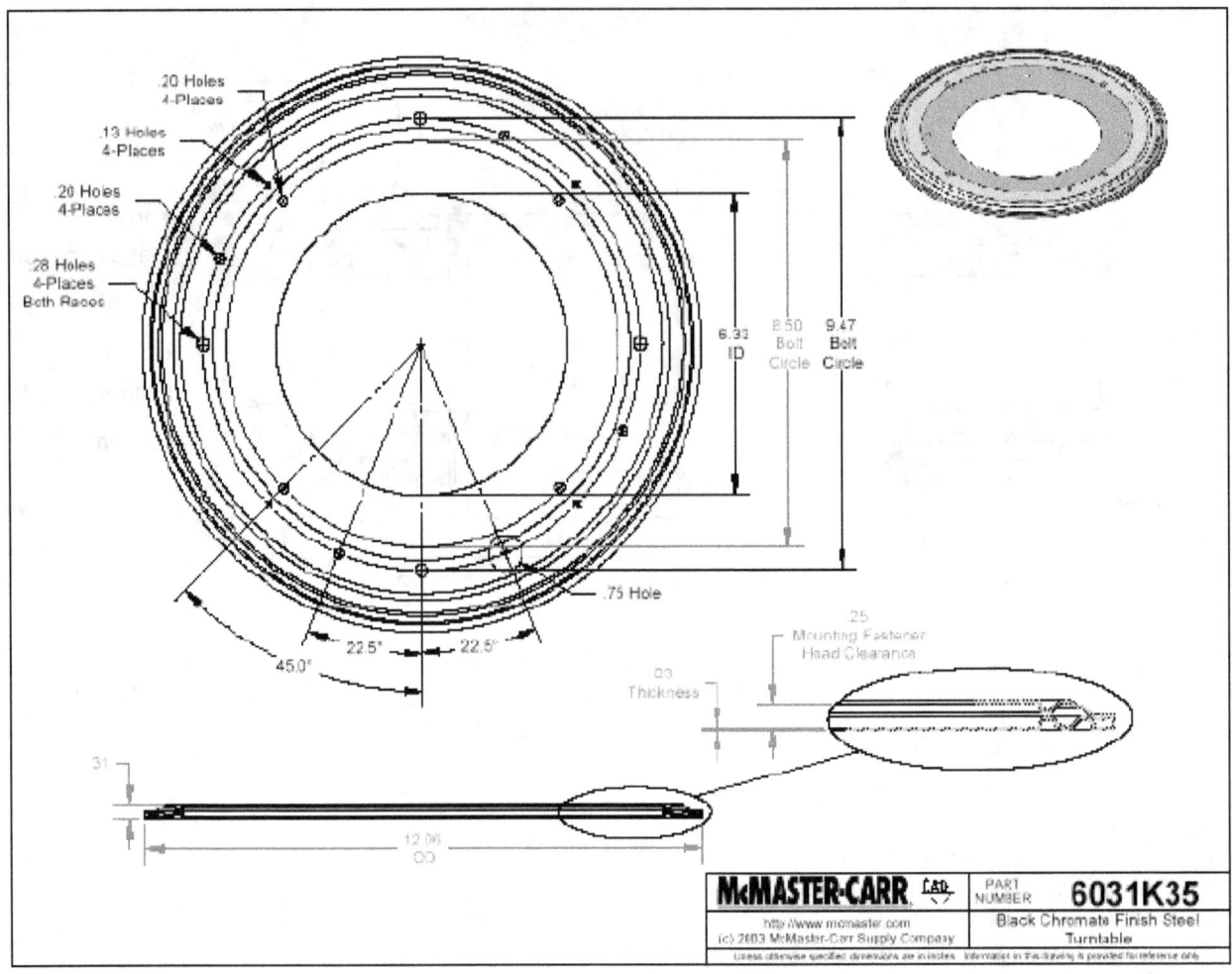

This drawing provided courtesy of McMaster-Carr. Other manufactured turntables could also be

Design of the HLPR Chair

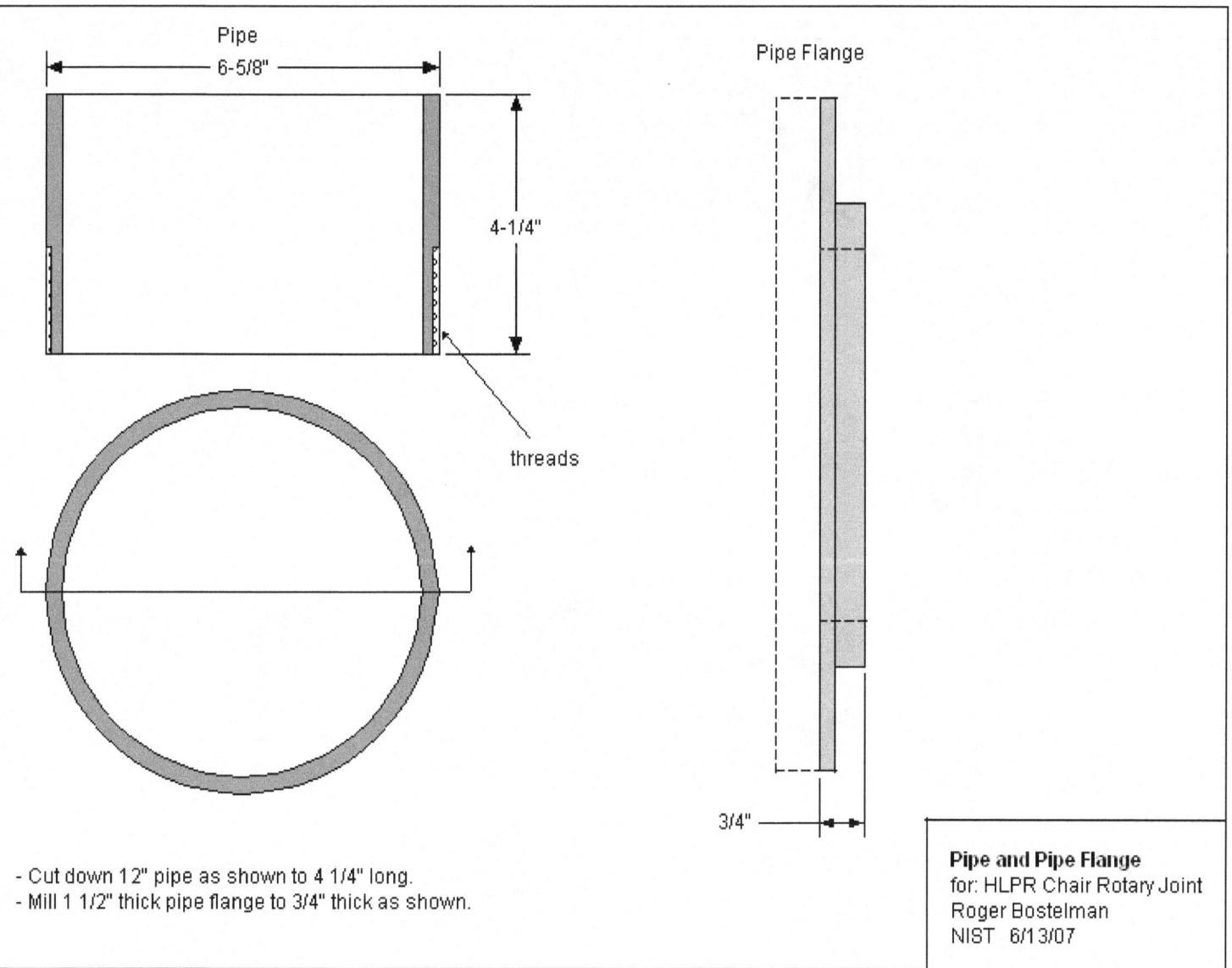

Design of the HLPR Chair

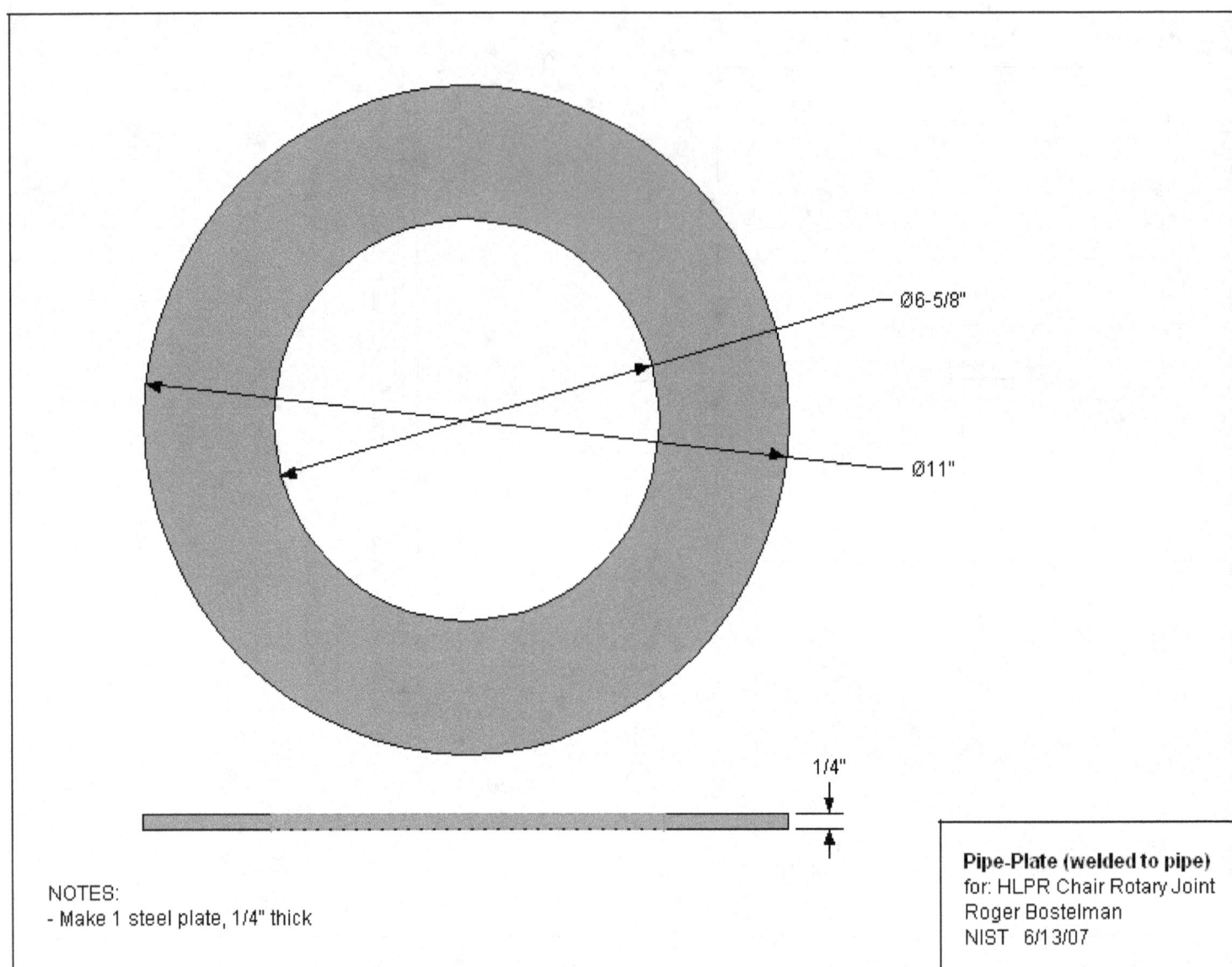

Design of the HLPR Chair

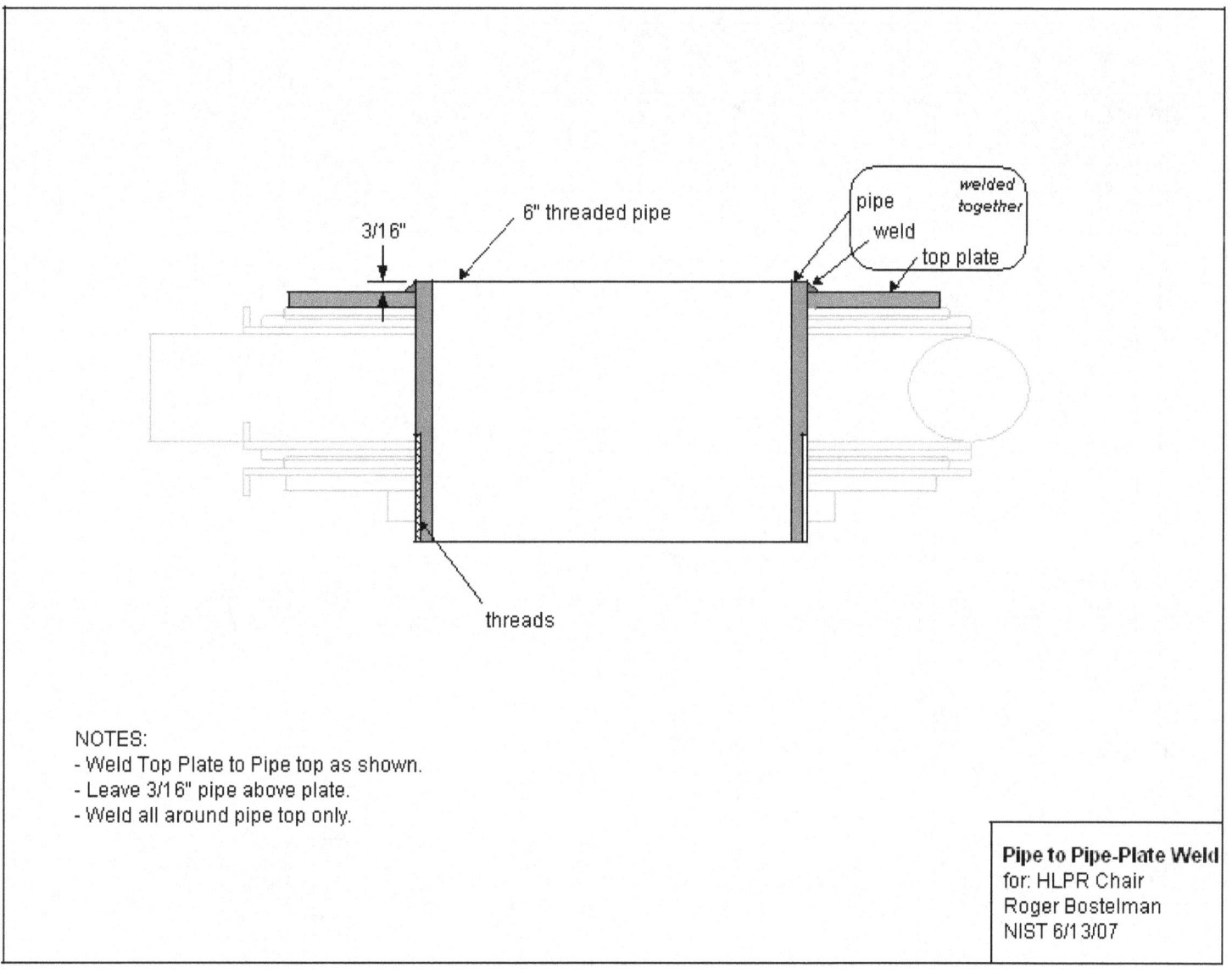

Design of the HLPR Chair

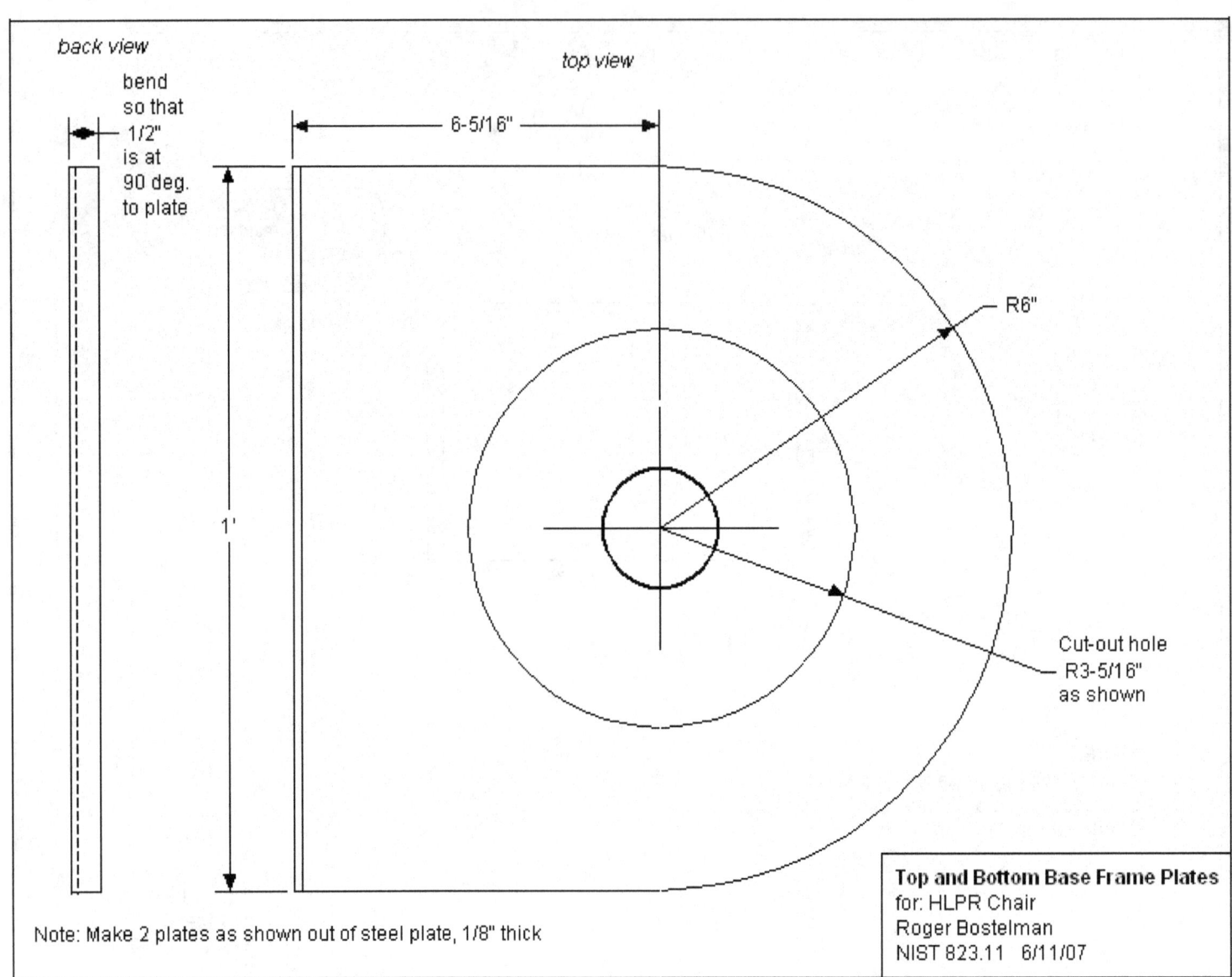

Design of the HLPR Chair

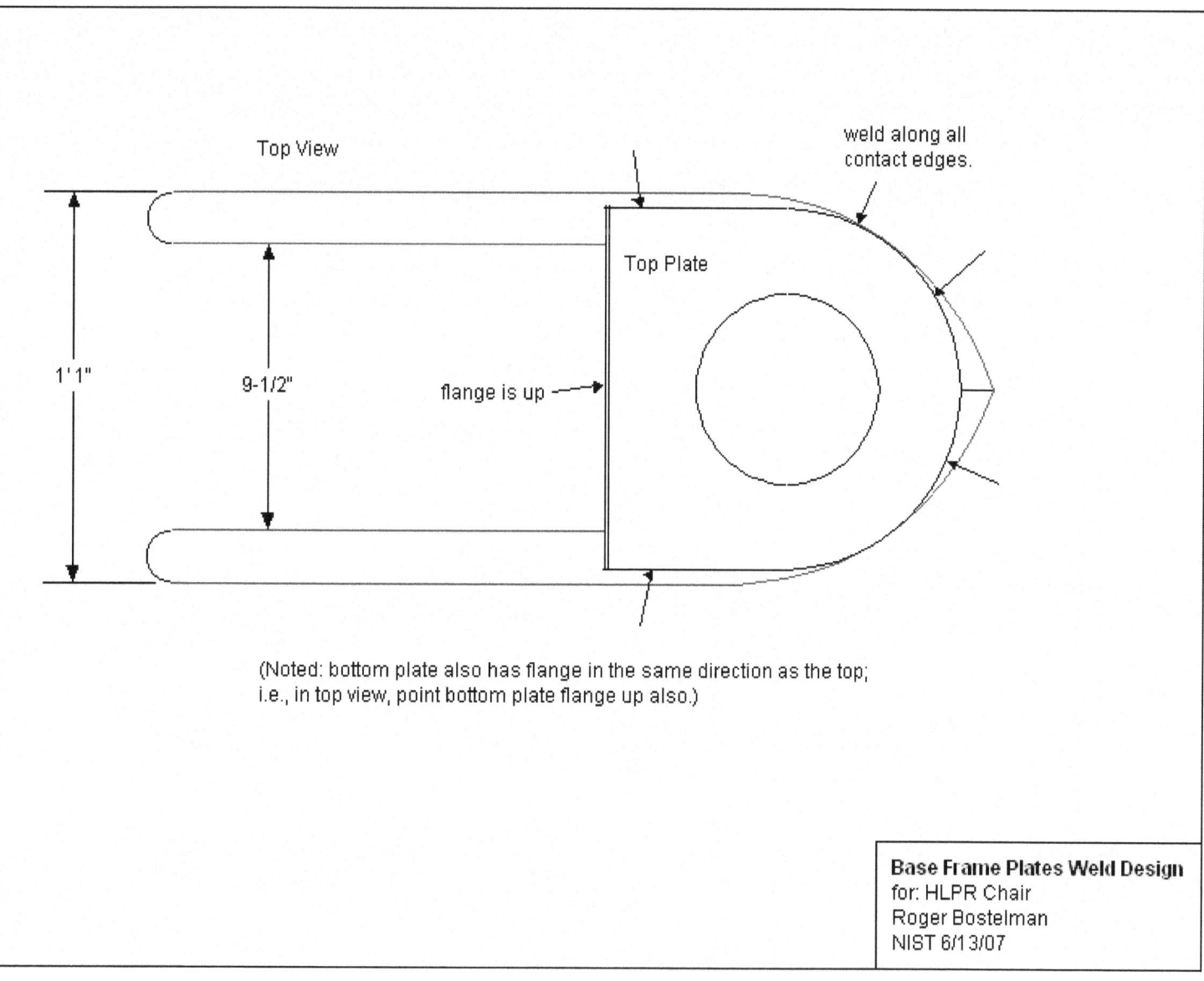

(Noted: bottom plate also has flange in the same direction as the top; i.e., in top view, point bottom plate flange up also.)

Base Frame Plates Weld Design
for: HLPR Chair
Roger Bostelman
NIST 6/13/07

Design of the HLPR Chair

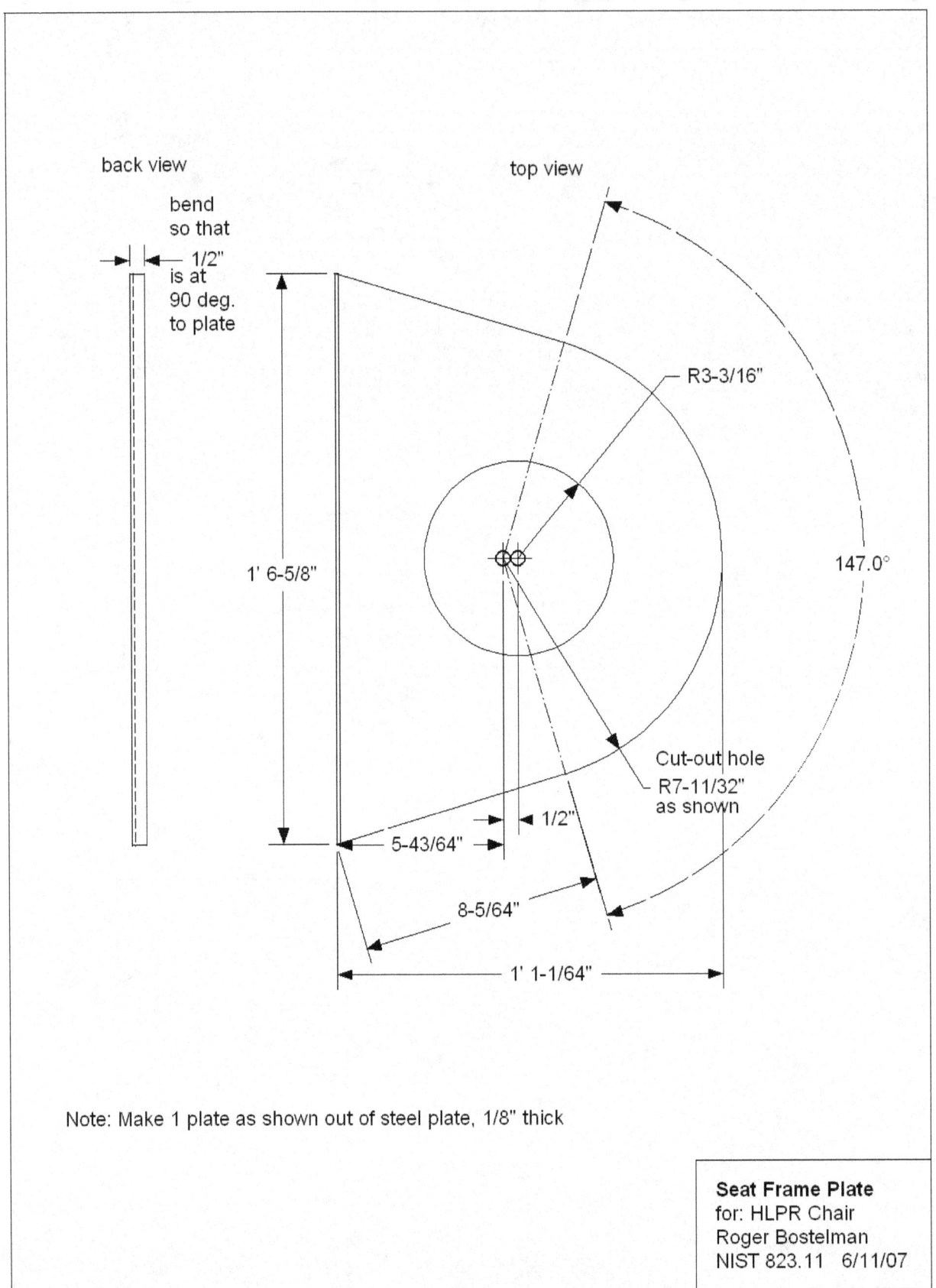

Design of the HLPR Chair

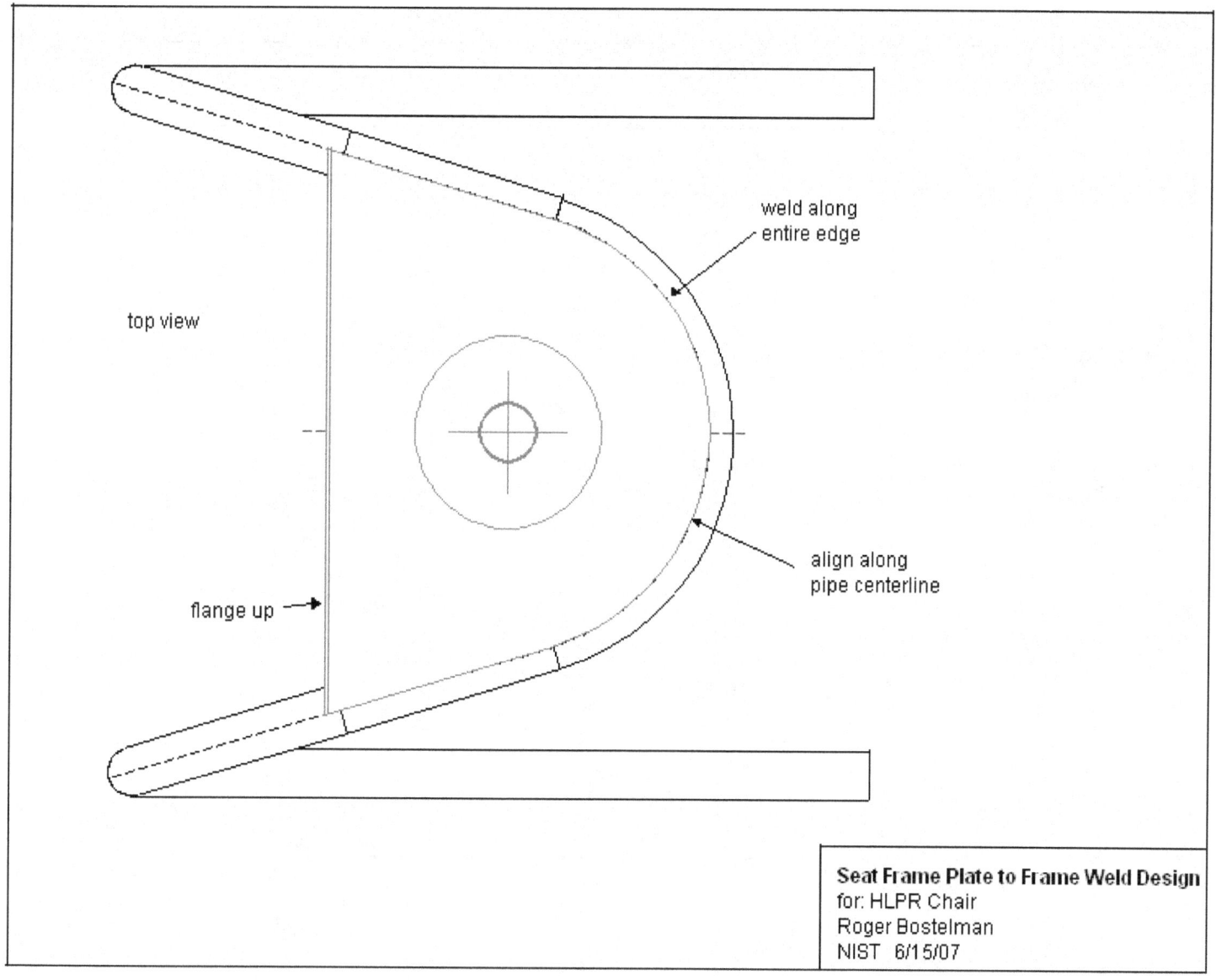

Design of the HLPR Chair

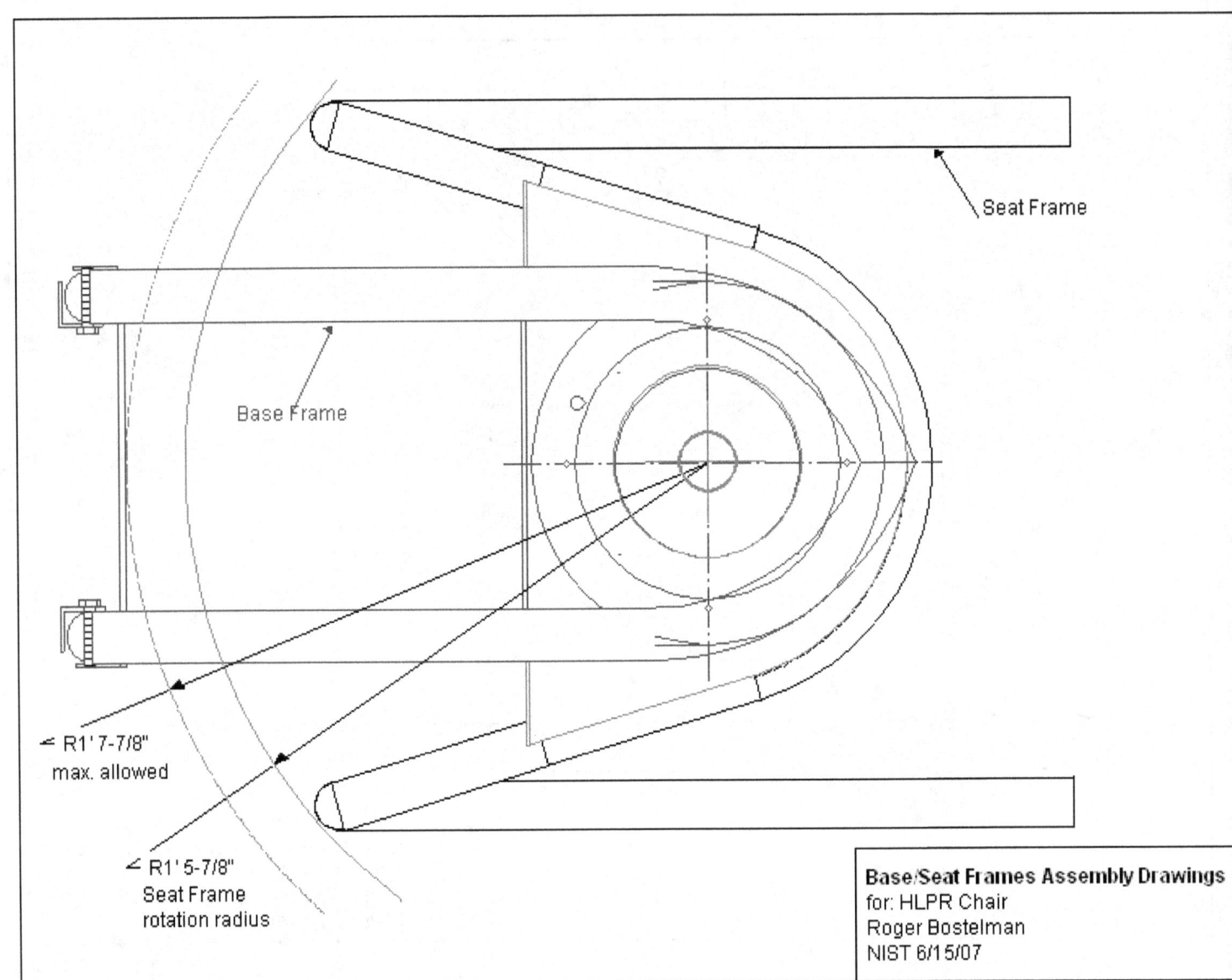

Design of the HLPR Chair

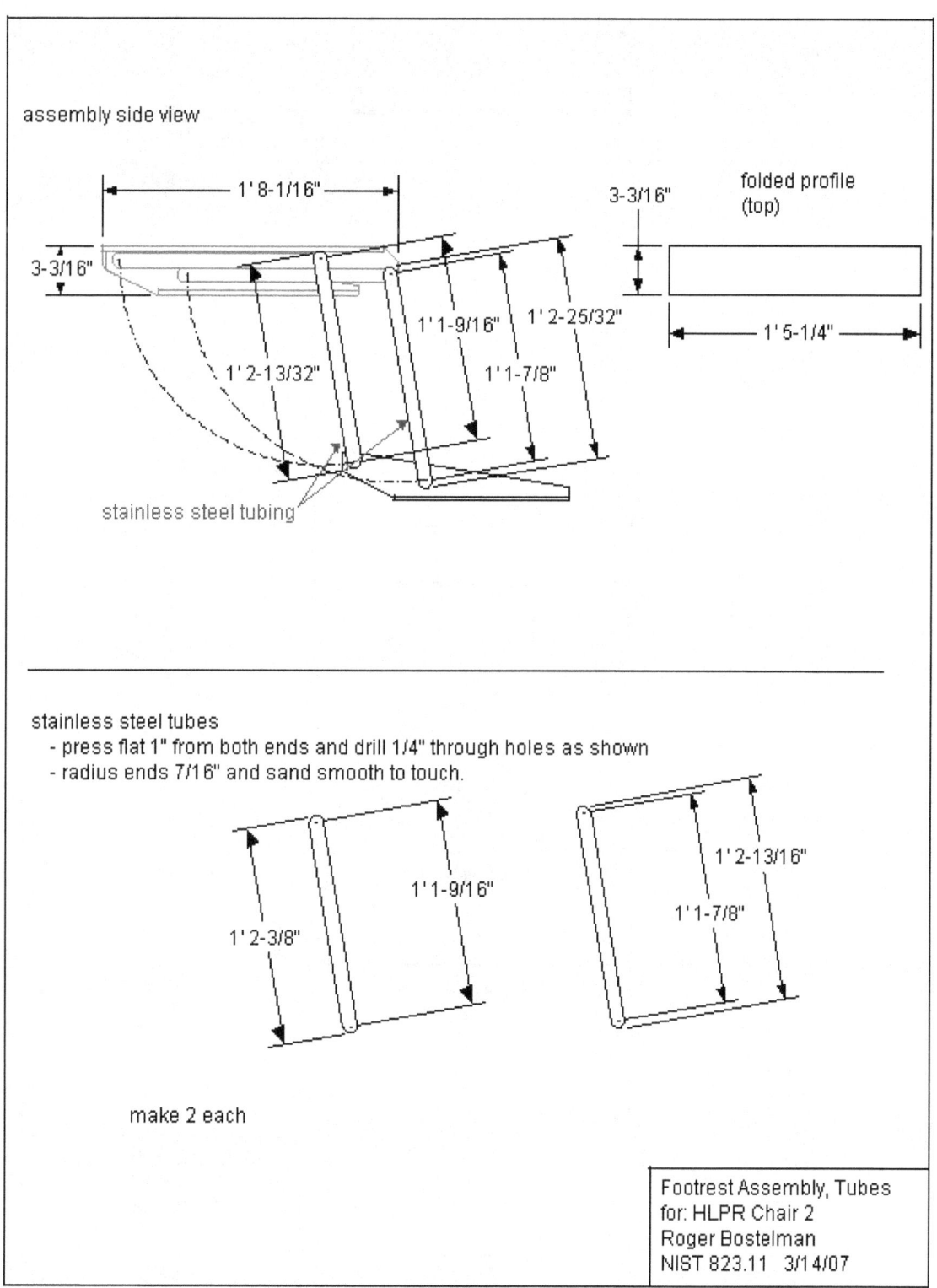

Design of the HLPR Chair

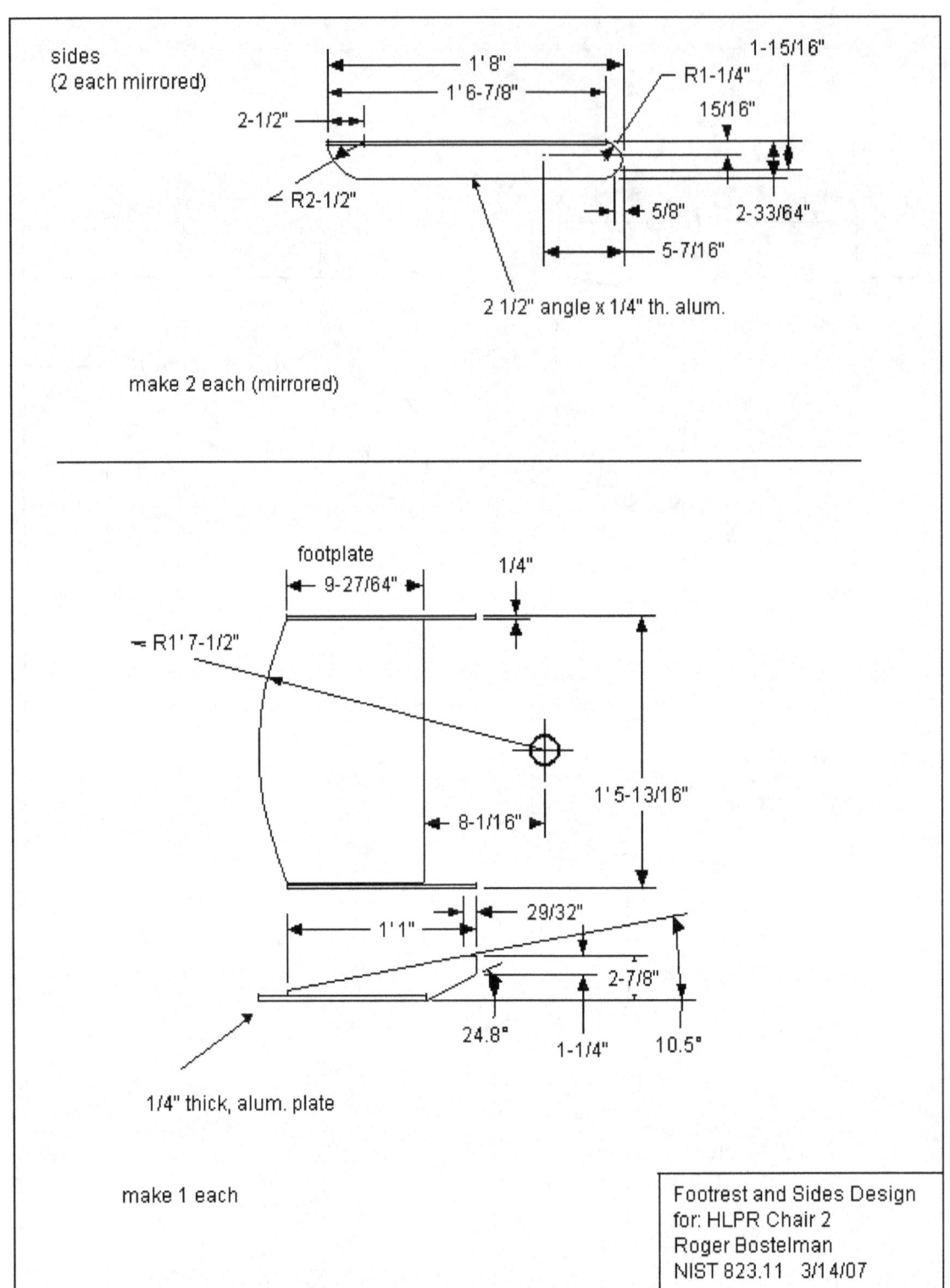

Design of the HLPR Chair

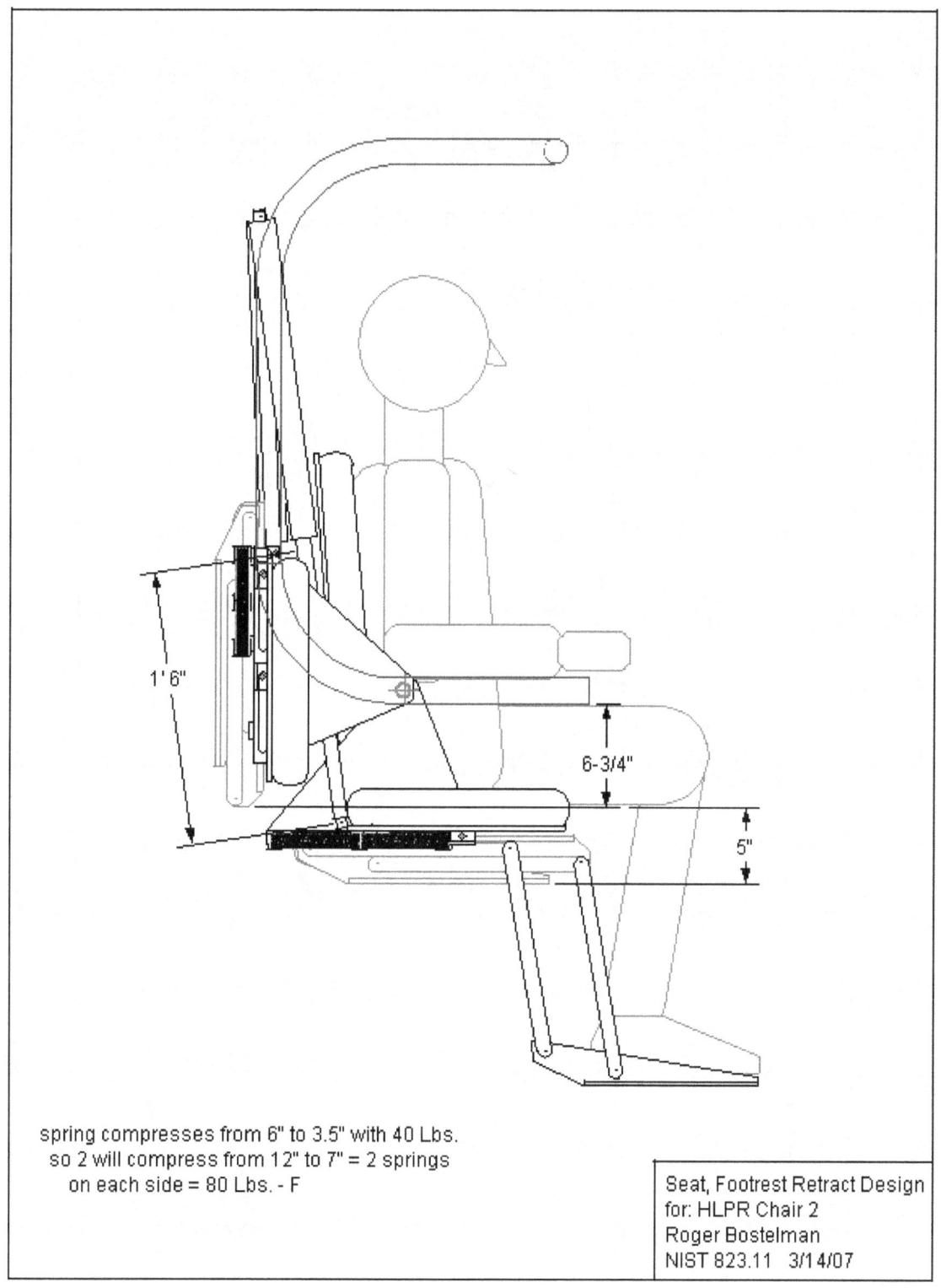

spring compresses from 6" to 3.5" with 40 Lbs.
so 2 will compress from 12" to 7" = 2 springs
on each side = 80 Lbs. - F

Seat, Footrest Retract Design
for: HLPR Chair 2
Roger Bostelman
NIST 823.11   3/14/07

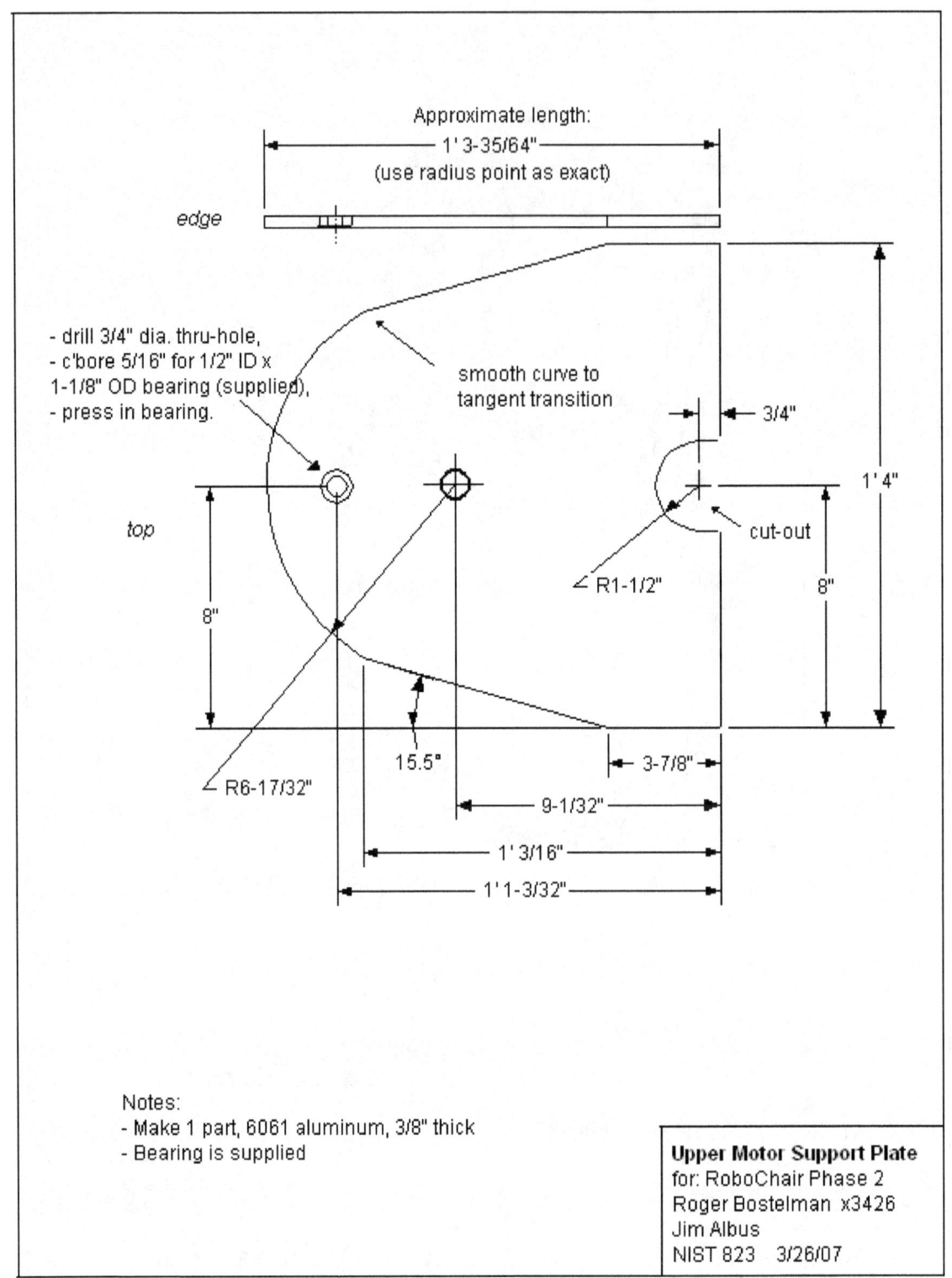

Design of the HLPR Chair

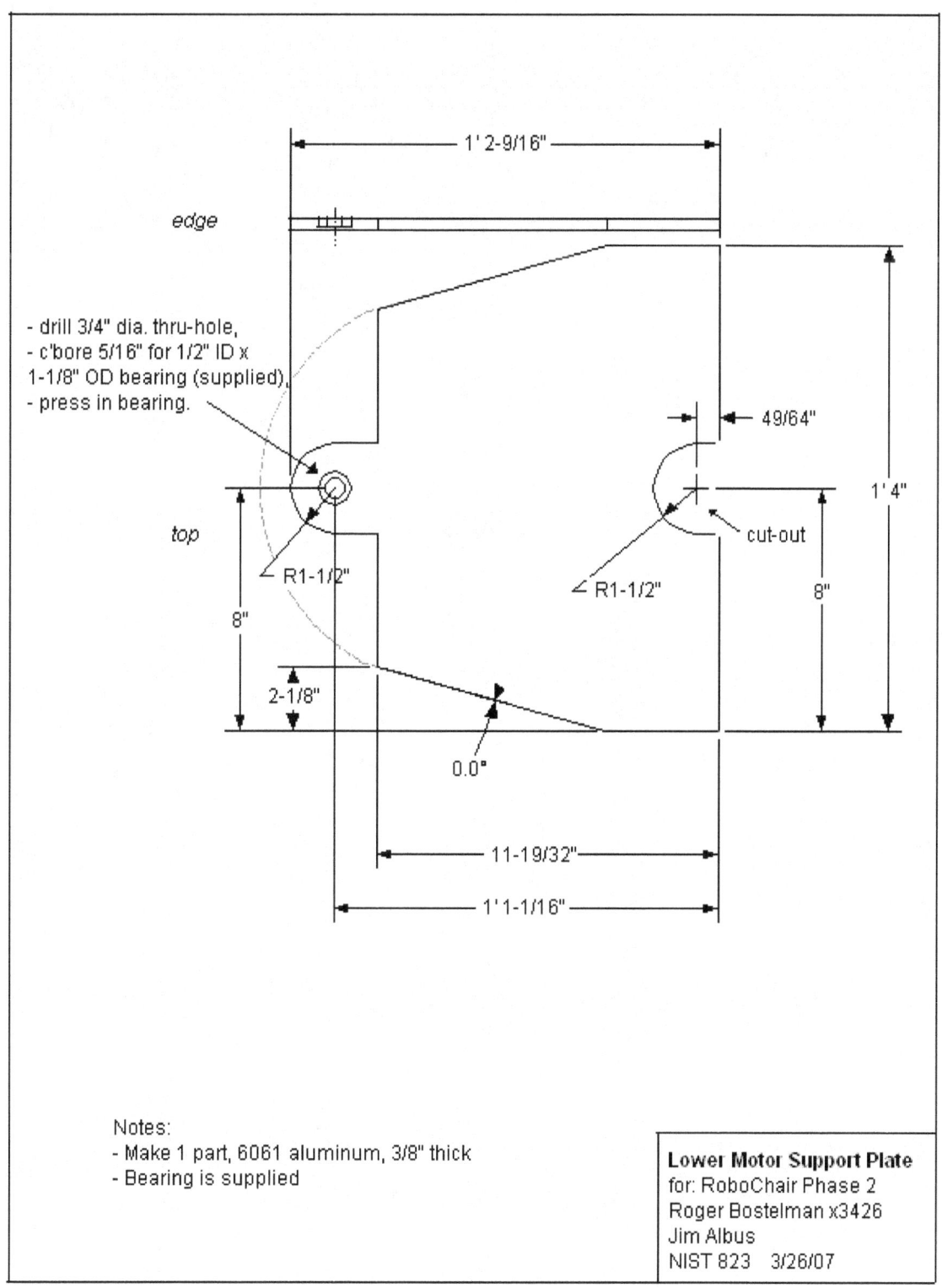

Electrical Design

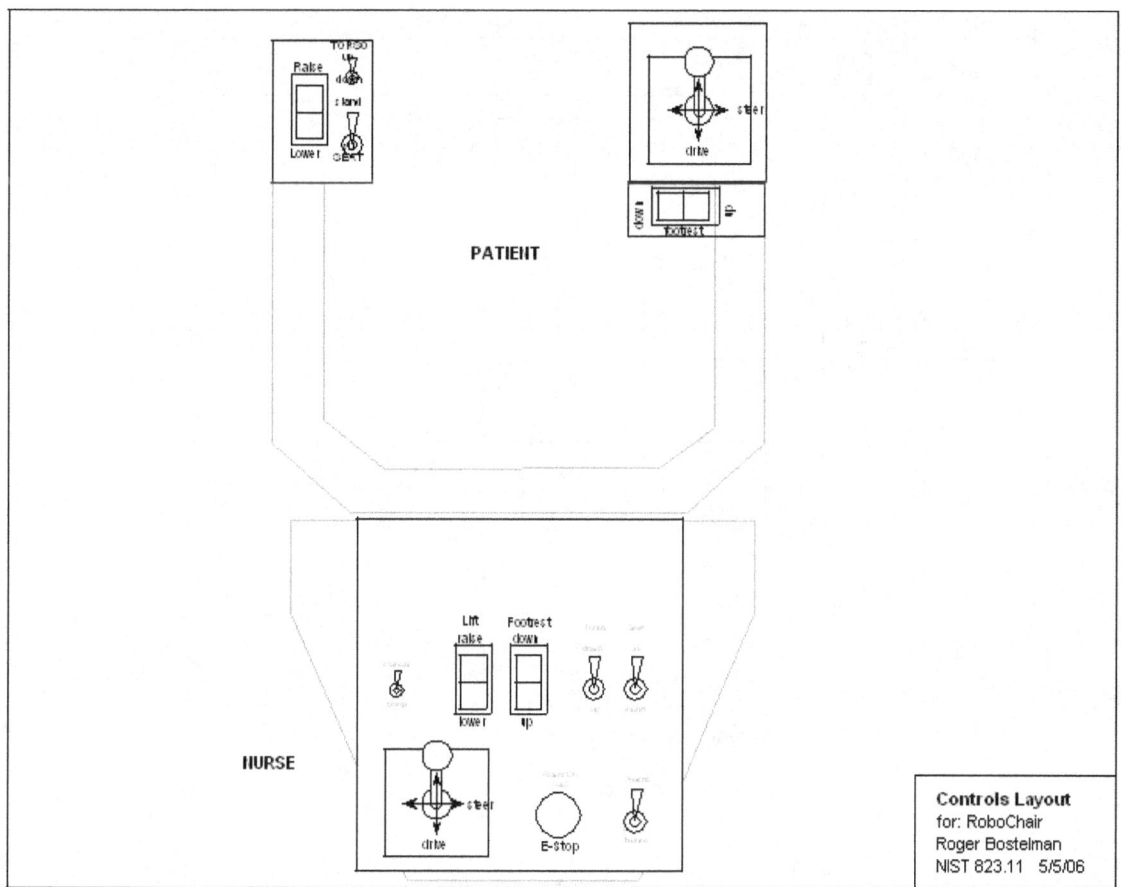

Design of the HLPR Chair

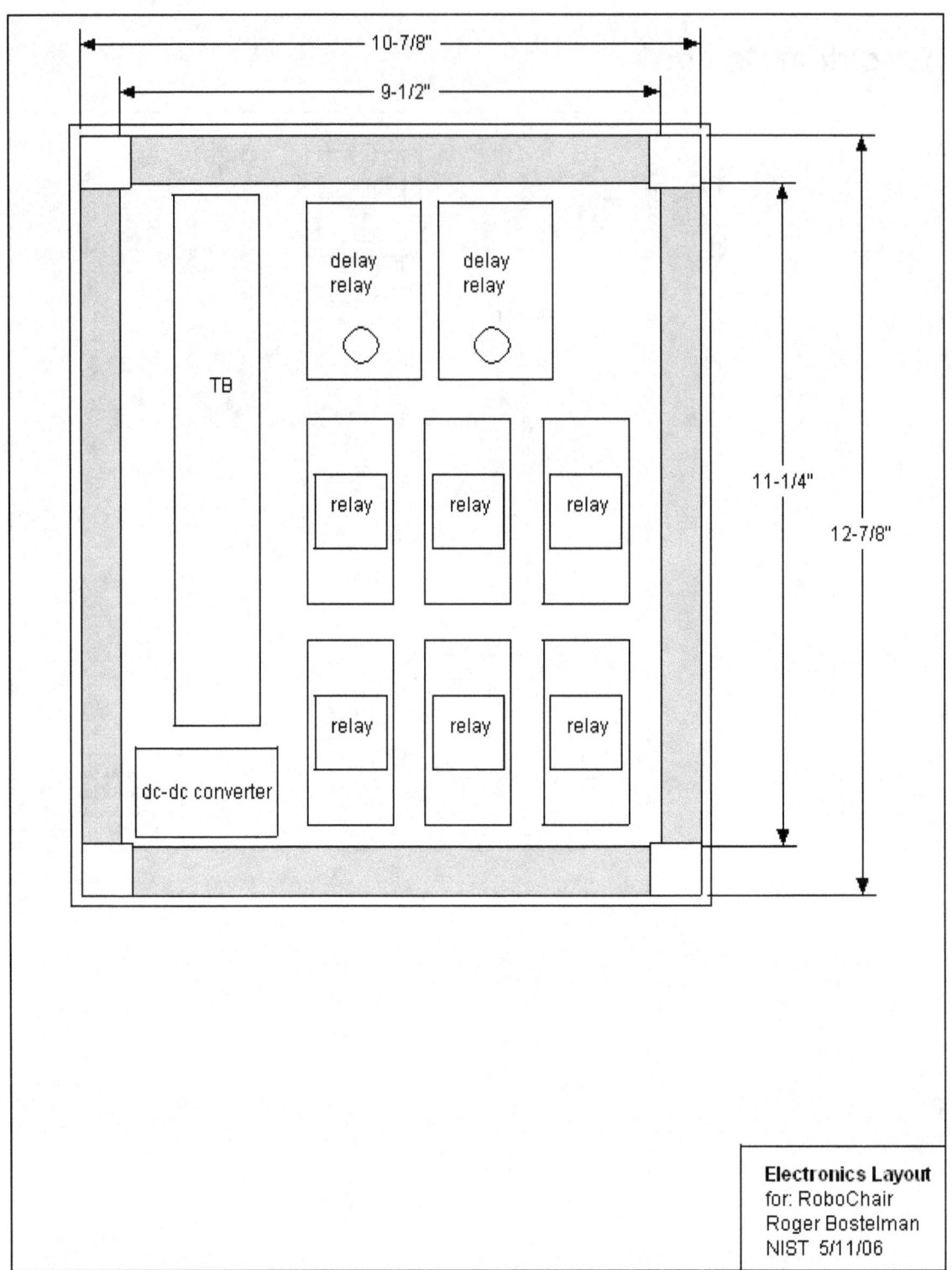

Design of the HLPR Chair

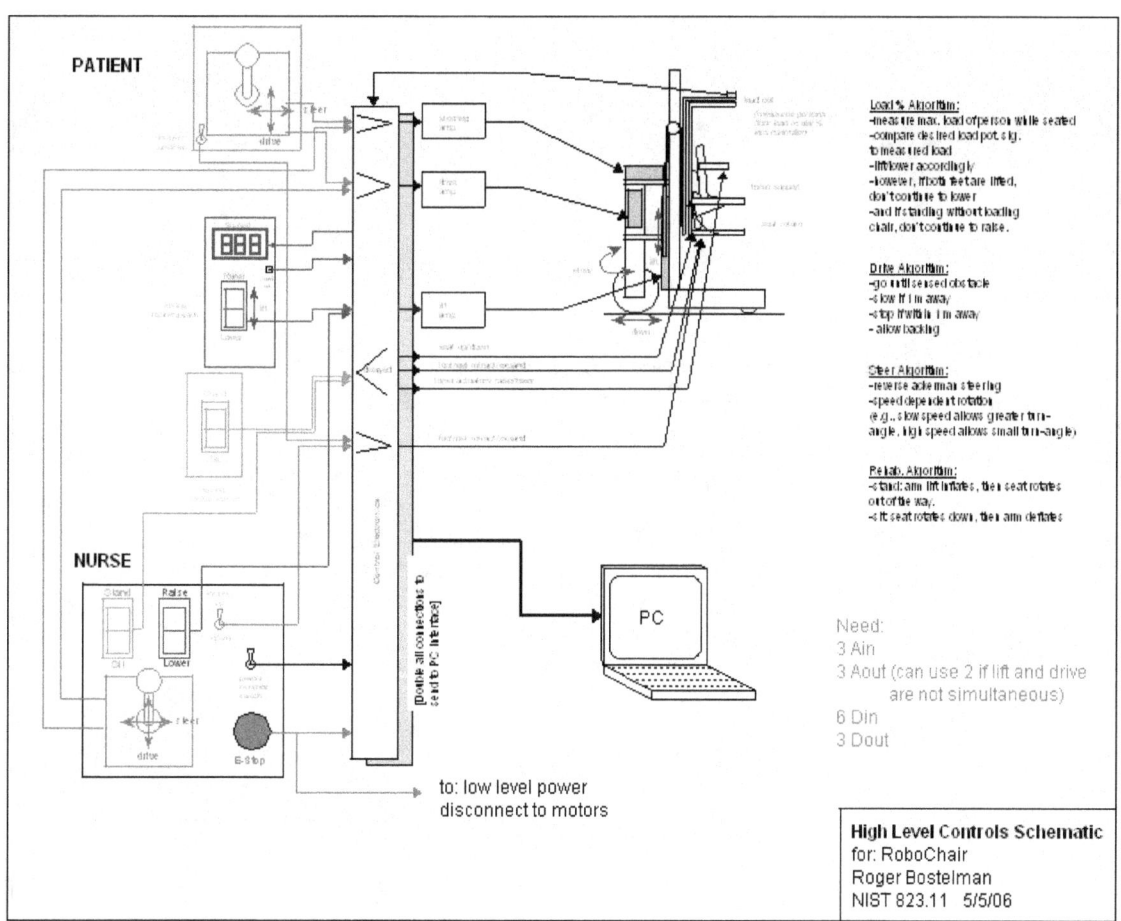

Design of the HLPR Chair

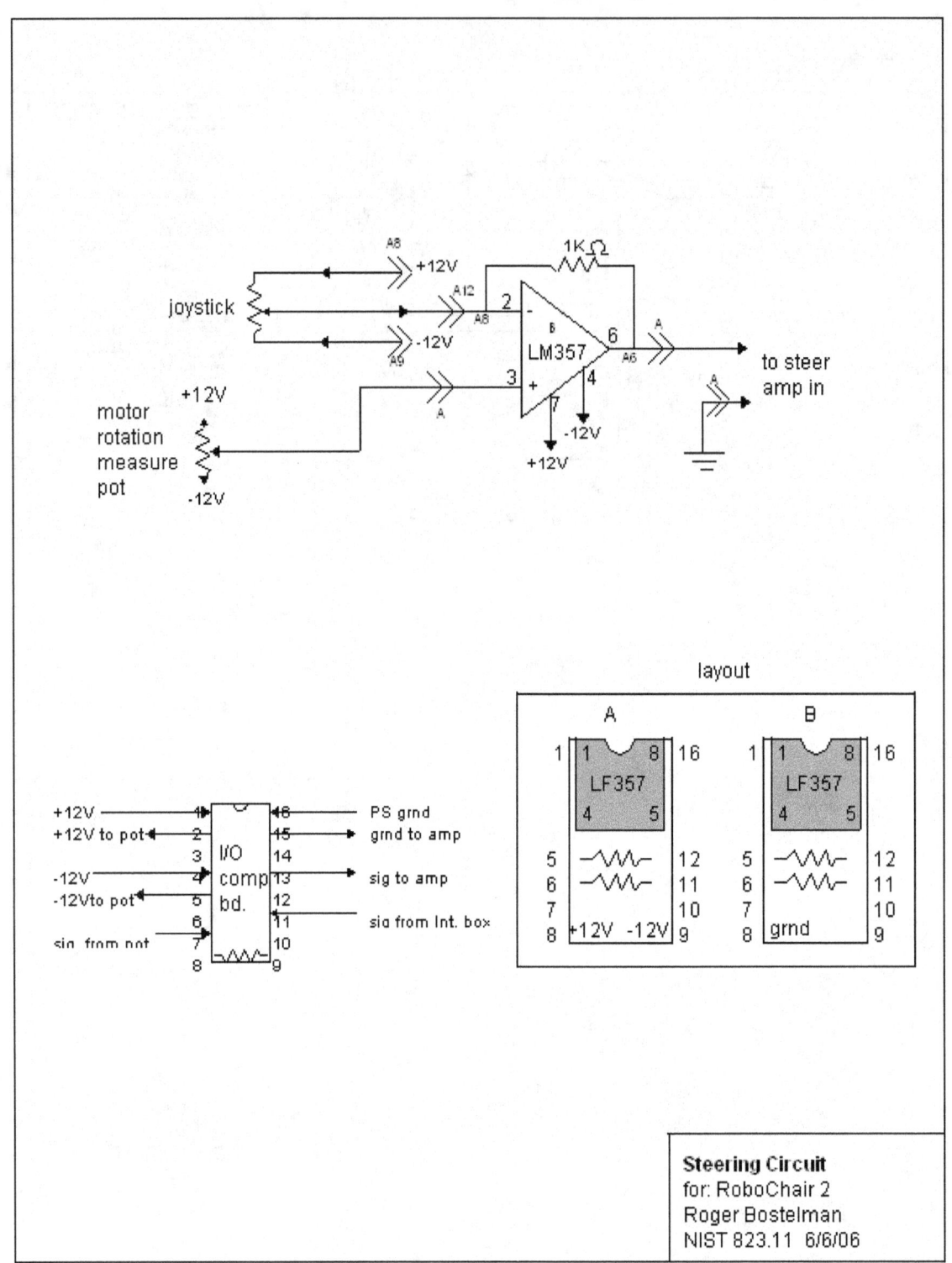

Steering Circuit
for: RoboChair 2
Roger Bostelman
NIST 823.11 6/6/06

Design of the HLPR Chair

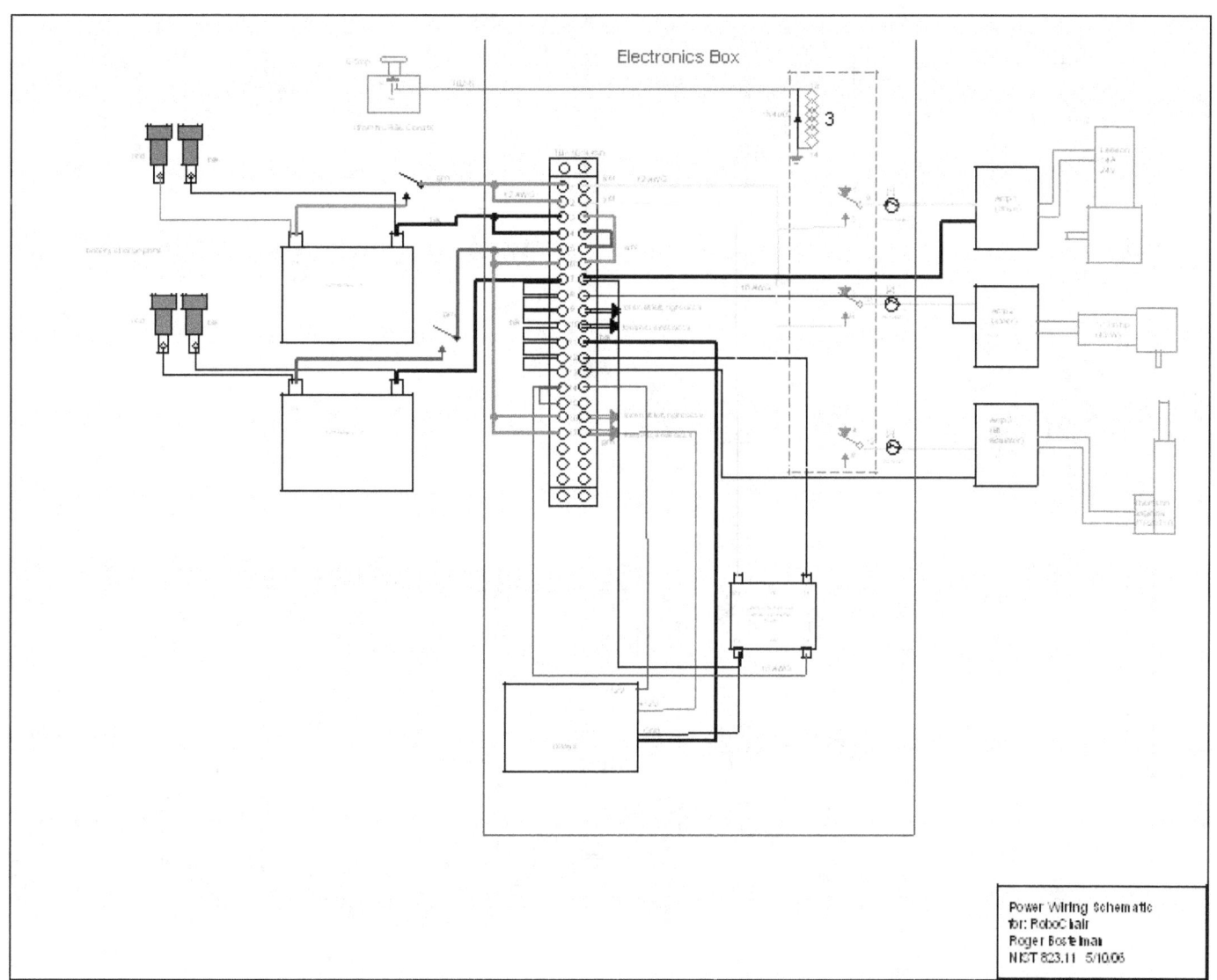

Design of the HLPR Chair

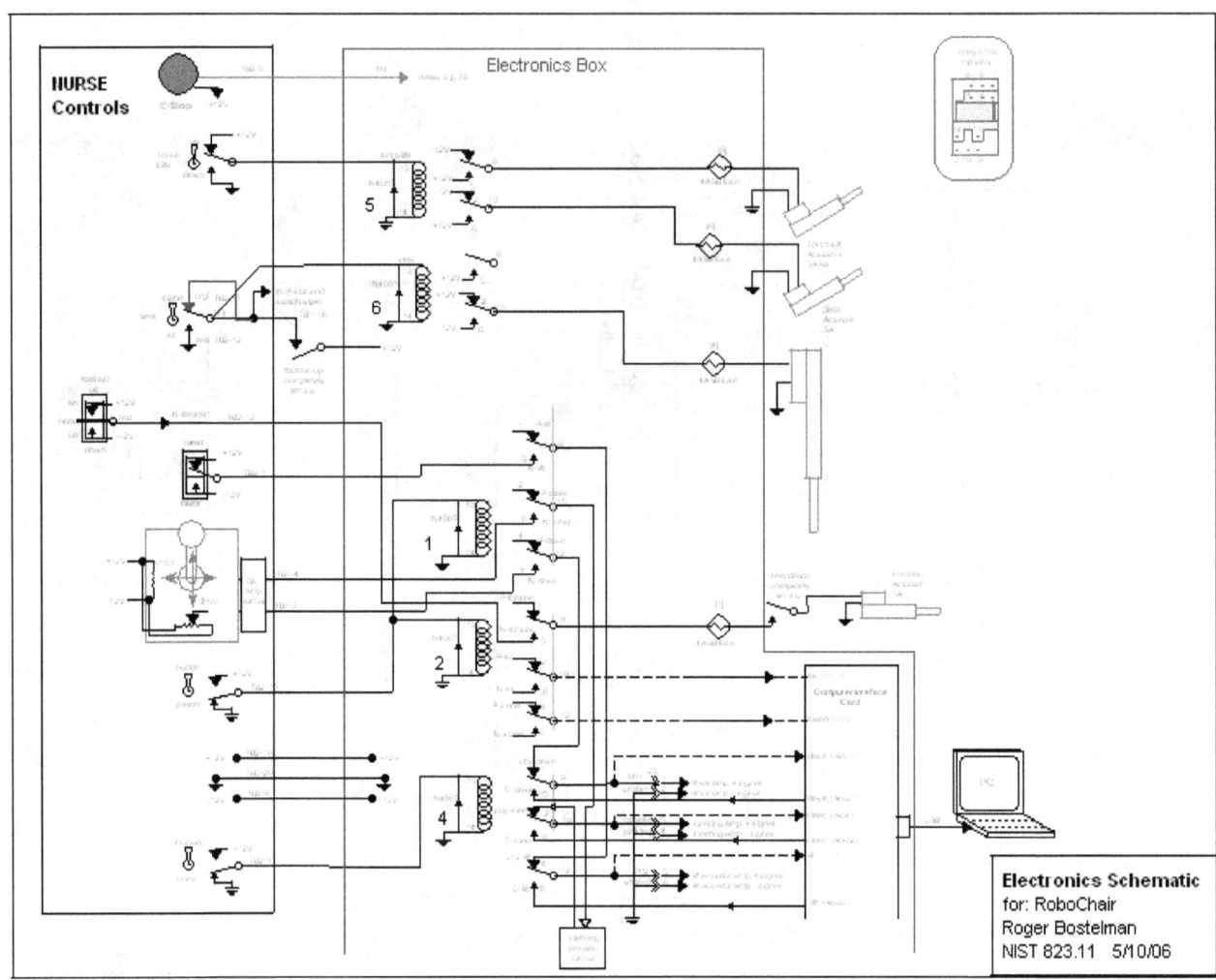

Design of the HLPR Chair

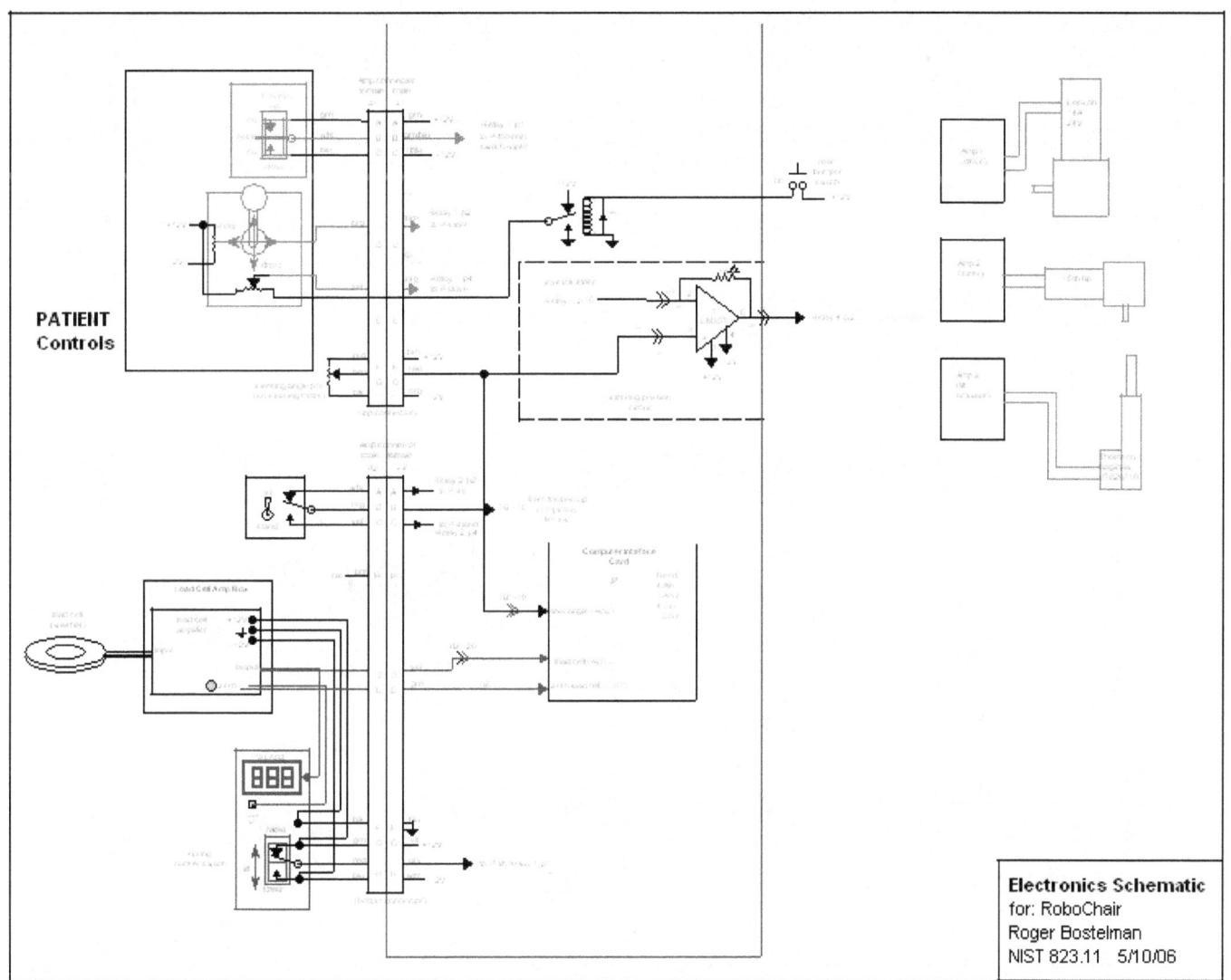

Design of the HLPR Chair

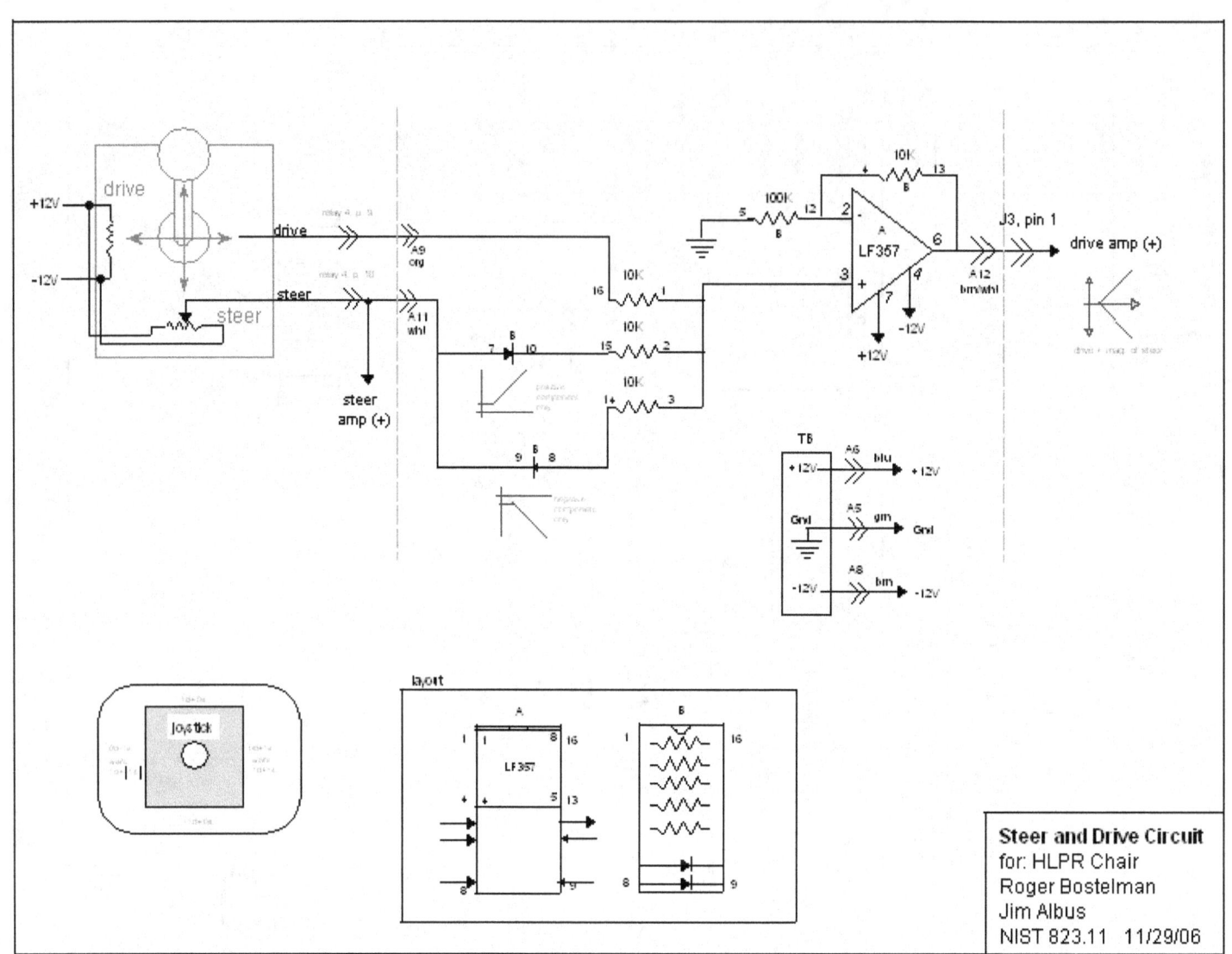

Design of the HLPR Chair

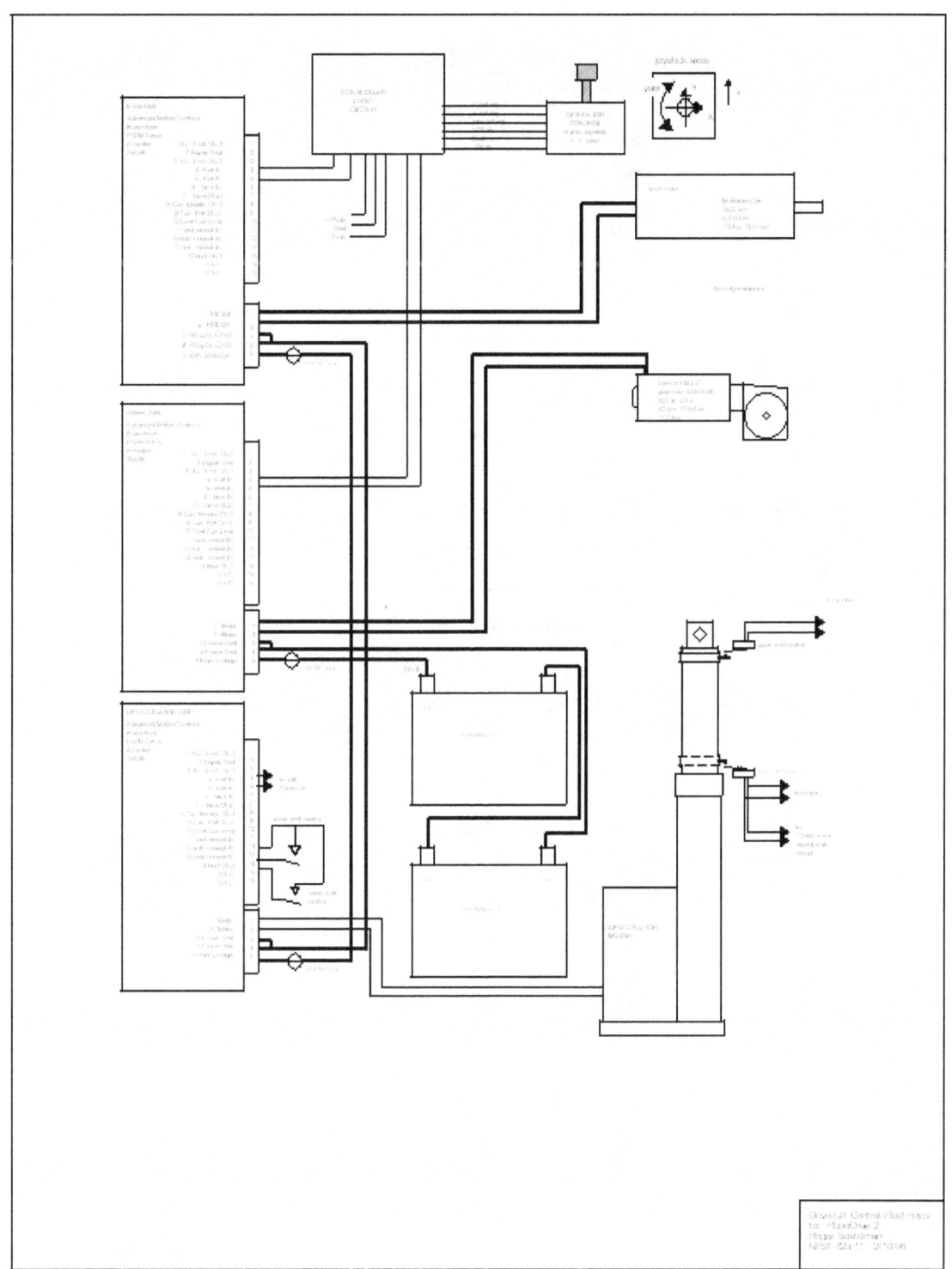

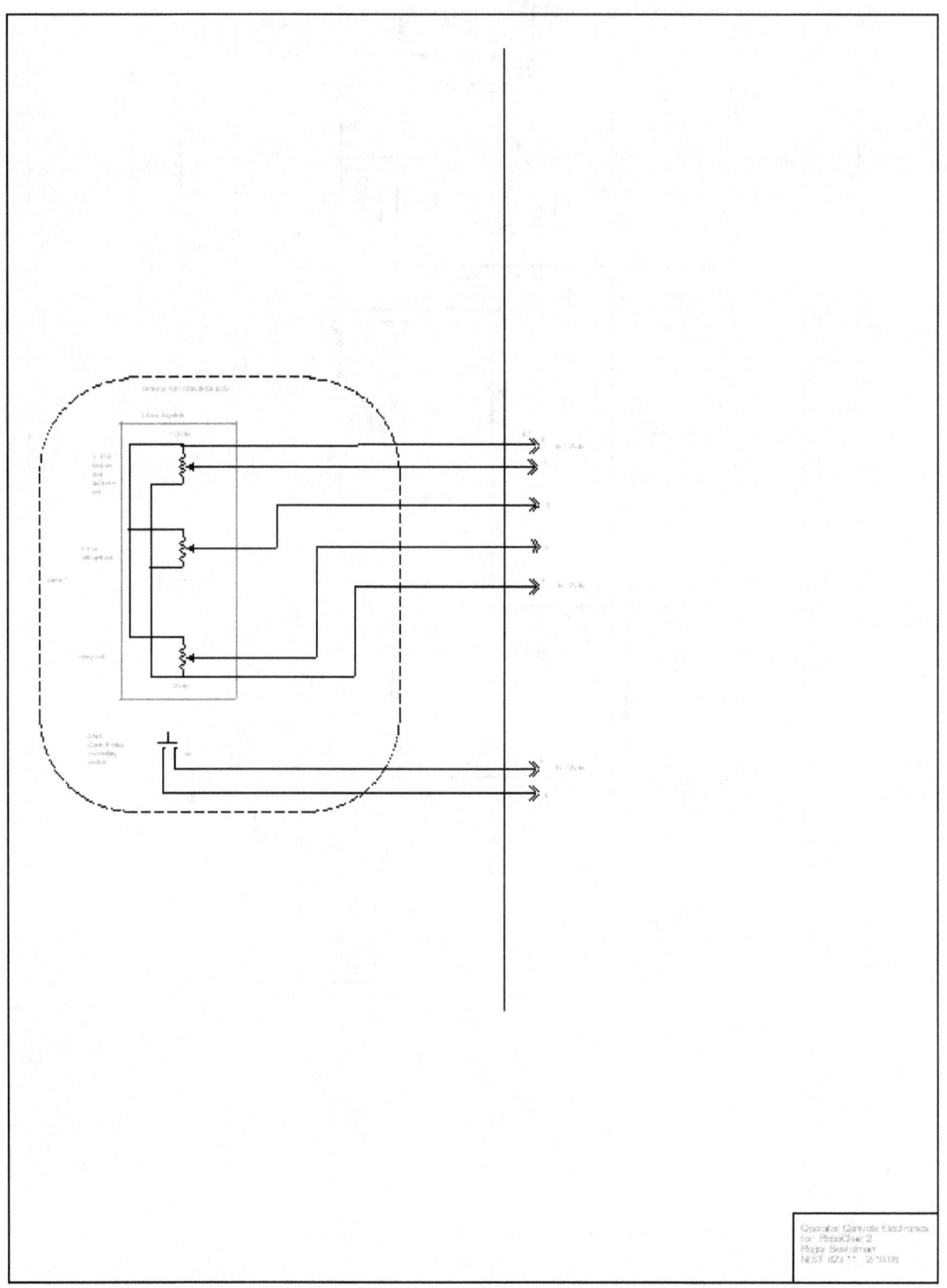

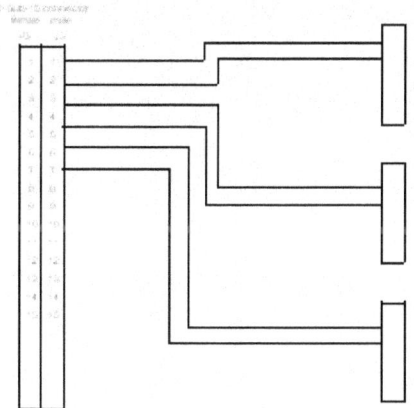

Connector Wiring
for: RoboChair
Roger Bostelman
NIST 823.11 6/2/06

Towards Autonomous Control

Refer to Photograph 8 for the following design drawings in this section.

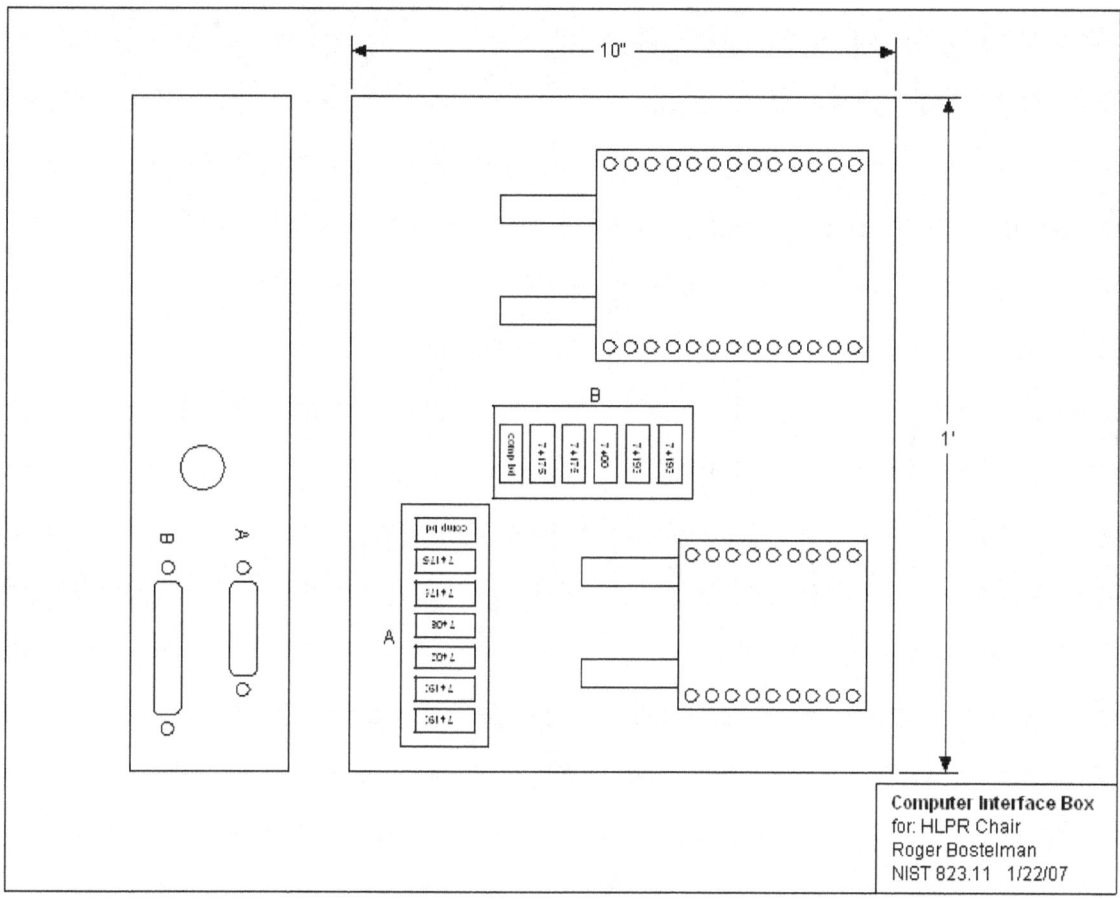

Design of the HLPR Chair

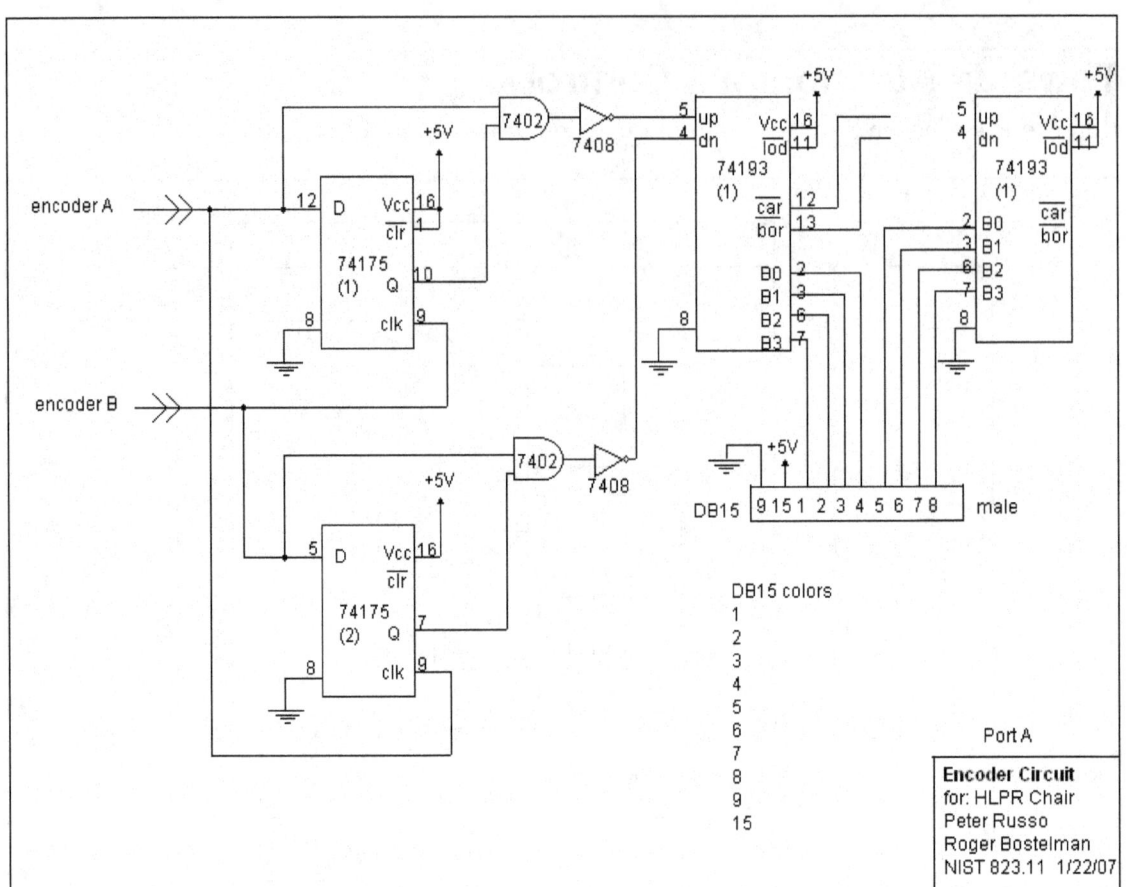

Design of the HLPR Chair

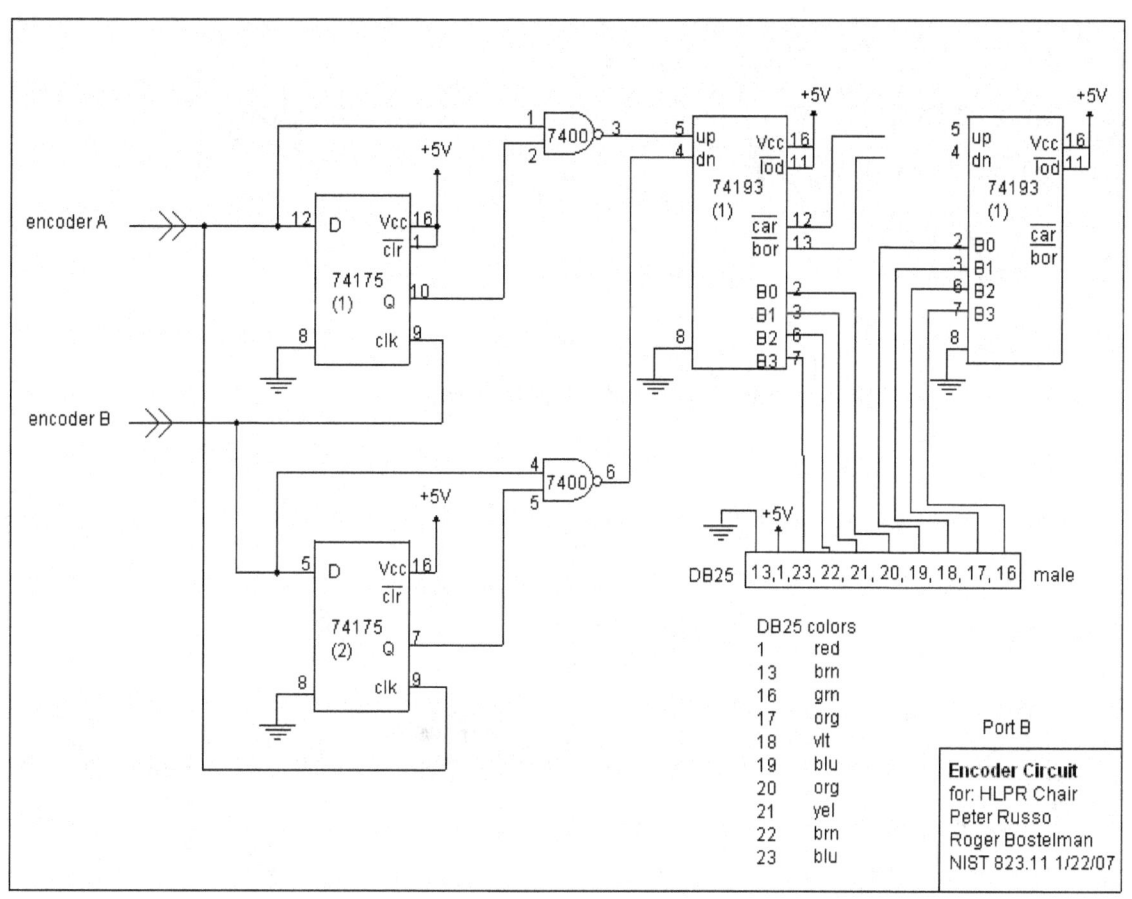

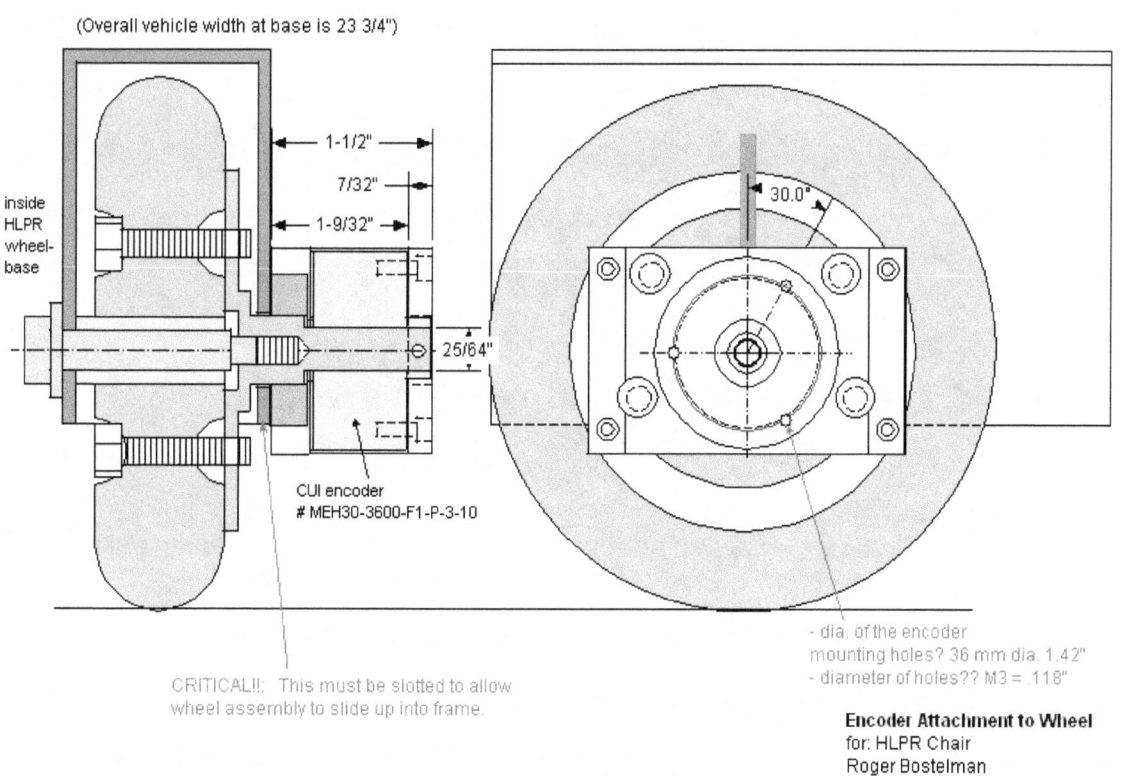

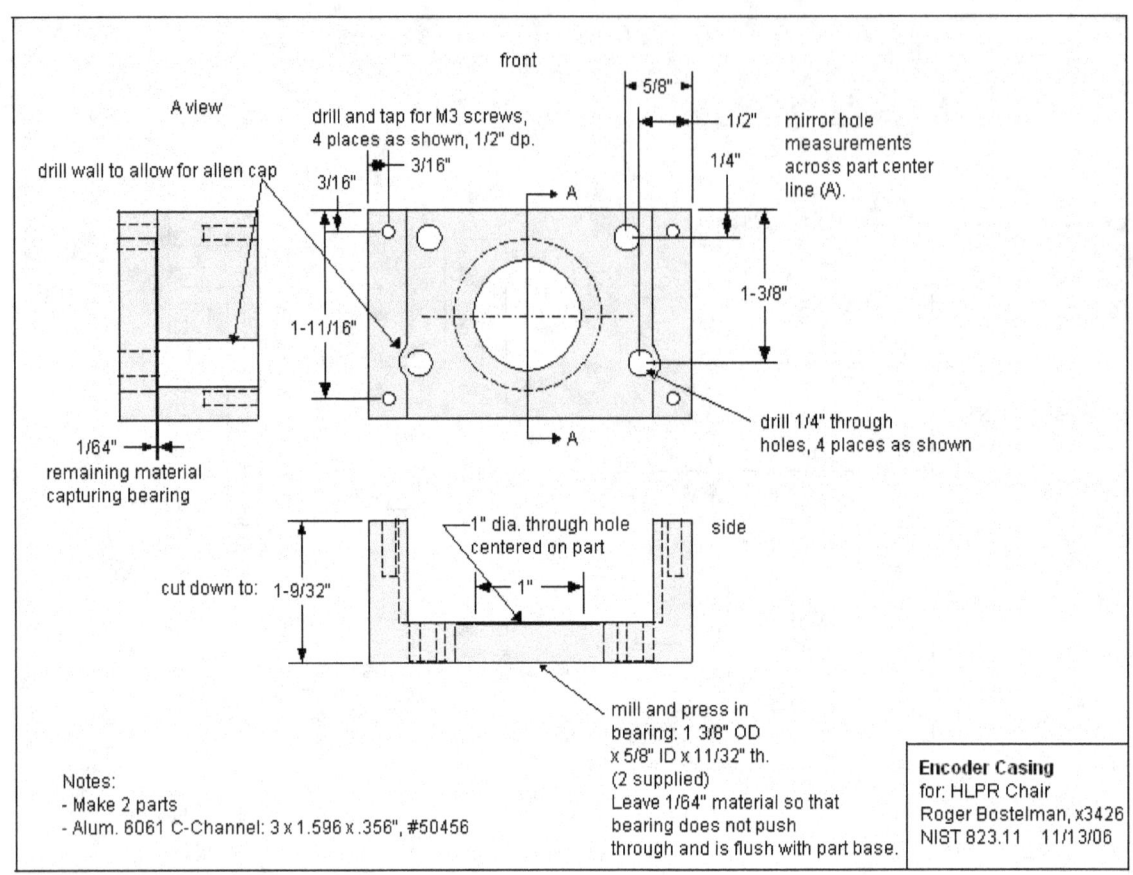

Design of the HLPR Chair

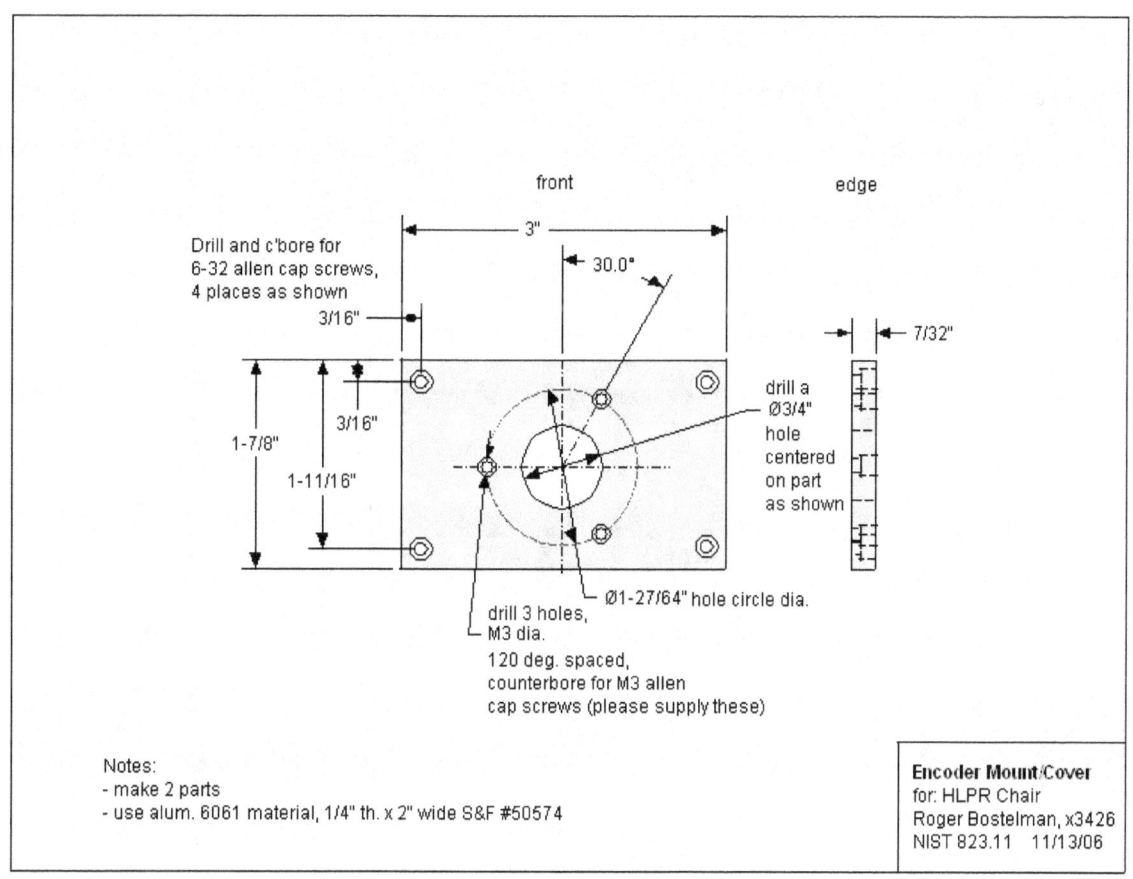

Design of the HLPR Chair

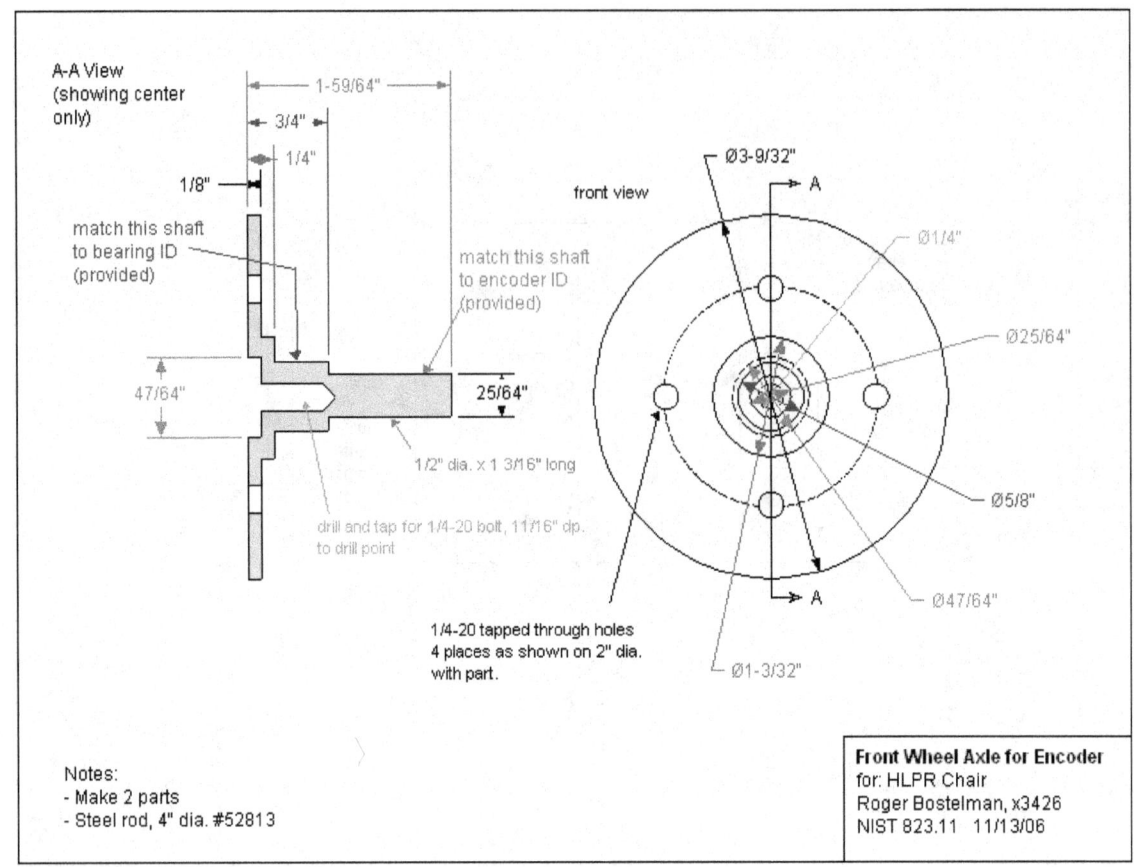

Design of the HLPR Chair

Stability Testing

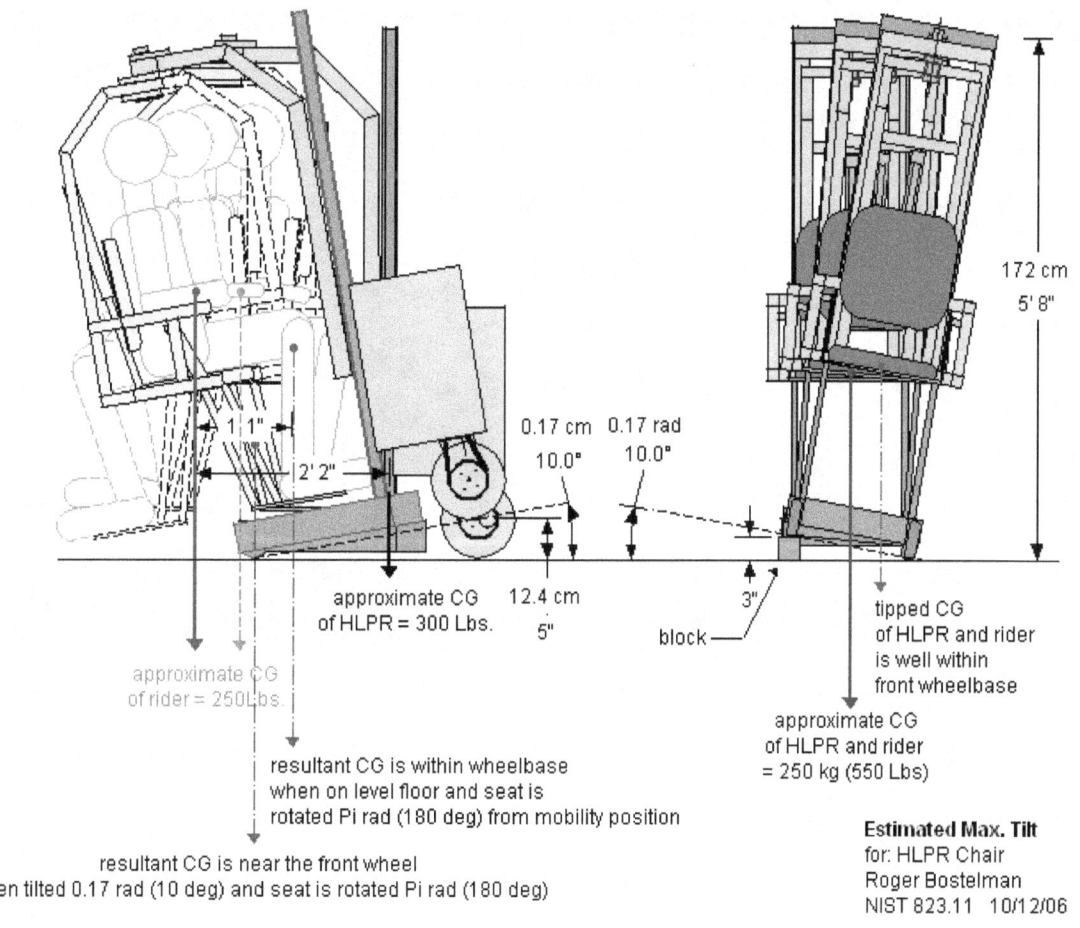

Estimated Max. Tilt
for: HLPR Chair
Roger Bostelman
NIST 823.11 10/12/06

Prototype Cost Estimate

Item	Manufacturer/Vendor	no. each	unit cost	cost
Lift Actuator		1	286.00	286.00
Drive motor, 14A, PM	Leeson	1	155.26	155.26
Steering motor, 1/8 hp	Leeson	1	285.00	285.00
Amplifiers Servo	Systems	3	295.00	885.00
Gear Reducers, couplings	McMaster Carr	2	321.60	643.20
Lubricant McMaster	Carr	1	27.70	27.70
Seat Actuator, 12"	McMaster Carr	1	294.74	294.74
Footrest Actuator, 3"	Firgelli	1	80.00	80.00
Torso Lift Actuator, 3"	Firgelli	2	80.00	160.00
Arm pads - foam, vinyl	JoAnn Fabrics	1	20.00	20.00
Seat, Backrest		1	50.00	50.00
Wheel, tire	McMaster Carr	1	33.00	33.00
chain (unit = per foot)	McMaster Carr	3	3.47	10.41
Sprockets, 1.67, 5.06"	McMaster Carr	1	38.33	38.33
Electrical boxes	Lowes	3	13.00	39.00
Wire		1	100.00	100.00
Switches		4	40.00	160.00
Circuit Breakers		5	40.00	200.00
Joystick		2	300.00	600.00
Misc. Elect. Parts		1	100.00	100.00
Fork Lift frame		1	500.00	500.00
Seatbelt		2	119.99	239.98
Batteries, 12V, PC925	Odyssey	2	100.00	200.00
Battery, 12V, 10Ah		1	25.35	25.35
Shops work**	NIST	1	5,000.00	5,000.00
Misc. Parts - bolts, etc.		1	200.00	200.00
		total:		$ 10,332.97

** Shop work for this prototype was dramatically reduced incorporating a redesigned seat and base frame through the use of bent tubing (see Modifications to Original Design section). The base frame was made from 5 cm (2 in) OD steel electrical conduit with a welded support bead and the seat frame was designed at NIST and procured from a pipe bending manufacturer. A 30 cm (12 in) diameter turntable bearing ring was used as the rotation device. The total for these components was approximately $400.00 verses $5,000.00.

www.ingramcontent.com/pod-product-compliance
Lightning Source LLC
Chambersburg PA
CBHW081731170526

45167CB00009B/3780